Routledge Revivals

Introductory Spatial Analysis

First published in 1981, *Introductory Spatial Analysis* uses ideas from dimensional analysis and stochastic process theory to provide a consistent, logical framework for map analysis. 'Geography is about maps', so the saying goes, yet there is no other textbook for geography students that combines the discussion of maps with a treatment of quantitative methods of map analysis. This book differs from most other quantitative or cartographic geography texts in three respects: first it is a geography, not a statistics book, and therefore problems are examined by looking at the types of data used and the varieties of maps drawn and then at the analytical procedures that may be used to detect significant spatial patterns; second, no attempt is made to introduce tests that treat data without reference to their spatial location; and third, no advice is offered on specifically cartographic questions of map drawing and design.

David Unwin's text will serve as a valuable introduction to the techniques of spatial analysis that are so important in contemporary geographical study.

Introductory Spatial Analysis

David Unwin

First published in 1981
by Methuen & Co Ltd.

This edition first published in 2024 by Routledge
4 Park Square, Milton Park, Abingdon, Oxon, OX14 4RN

and by Routledge
605 Third Avenue, New York, NY 10017

Routledge is an imprint of the Taylor & Francis Group, an informa business

Publisher's Note
The publisher has gone to great lengths to ensure the quality of this reprint but points out that some imperfections in the original copies may be apparent.

Disclaimer
The publisher has made every effort to trace copyright holders and welcomes correspondence from those they have been unable to contact.

A Library of Congress record exists under ISBN: 0416722008

ISBN: 978-1-032-78534-9 (hbk)
ISBN: 978-1-003-48835-4 (ebk)
ISBN: 978-1-032-78536-3 (pbk)

Book DOI 10.4324/9781003488354

INTRODUCTORY SPATIAL ANALYSIS

DAVID UNWIN

METHUEN

LONDON AND NEW YORK

First published in 1981 by
Methuen & Co. Ltd
11 New Fetter Lane, London EC4P 4EE

Published in the USA by
Methuen & Co.
in association with Methuen, Inc.
733 Third Avenue, New York, NY 10017

Printed in Great Britain by
Richard Clay (The Chaucer Press) Ltd
Bungay, Suffolk

British Library Cataloguing in Publication Data

Unwin, David
Introductory spatial analysis.
1. Geography 2. Spatial analysis (Statistics)
I. Title
910'.01 G128

ISBN 0-416-72190-7
ISBN 0-416-72200-8 Pbk (University paperback 748)

·CONTENTS·

·LIST OF FIGURES·

·LIST OF TABLES·

·NOTATION·

In an introductory text one has to steer a course between two extremes. Over-rigorous adherence to a particular notation scheme can just as easily mystify the reader as slackness can confuse. I have tried to make consistent use of lower case, subscripted as appropriate, for general quantities, reserving upper case for specific measures. Statistical population parameters are seldom used, so notation is confined to the Roman form for estimates. Greek symbols have been used where such use is common.

a = any area a_c = area of a circle
b = any exponent B = bifurcation ratio
c = any constant C = constant of path maintenance
C_m, C_a = coefficients of areal correspondence
$\mathbf{C}$ = a connectivity matrix
d = any density D = path density
D_I = index of dissimilarity
d_s = standard distance
d_{max}, d_{crit} = test and critical value in a Kolmogorov-Smirnov test
$df1$, $df2$ = degrees of freedom in analysis of variance
e = 2.71828 . . . , or in context the number of paths in a network, or the residual from a regression or trend surface
E = choropleth error index
f = any frequency F = path frequency
f( ) denotes 'some function'
g = number of subgraphs in a topological graph
j = count of joins J = total count of joins
k = any constant, number of quadrats, colours, or contours on a map
l = any length subscripted by r (radius), e (expected), ri (radius of

of internal circle), *re* (radius of external circle) 1 (long axis), 2 (short axis), *q* (link length), *ic* (distance of point to mean centre) and *a*, *b*, *c* (sides of a triangle)

L = the dimension of length

m = a constant number, number of points in a quadrat

M = the dimension of mass

M = any matrix of numbers

n = number of cases

n_q = contact number

N = total number

p,q = probabilities

P = population

r = product moment correlation coefficient

R = nearest-neighbour index

s = standard deviation

S = sinuosity ratio

S_2 = compactness ratio

t = Student's *t*-index for comparing two means

T = the dimension of time

v = vector distance with components v_e, v_n and resultant v_r; in context the number of nodes

x, y, z = locational and value co-ordinates; in context z also indicates a standard normal deviate

BB, WW and BW = counts of joins in autocorrelation tests

α = alpha index in network analysis

β = beta index in network analysis

γ = gamma index in network analysis

θ = theta; any angle

λ = lambda, the product Np in quadrat analysis

μ = mu, the Cyclomatic number in network analysis

π = pi, the value 3.1416

∇ = del operator, sometimes 'grad' or 'nabla'; the gradient of a scalar field

·ACKNOWLEDGEMENT·

Most of the first draft of this book was written whilst on study leave in Ballater, Scotland. I am grateful to the University of Leicester for making this possible and to my wife Kathy for keeping me writing in the midst of so many tempting hills. Further thanks are also due to Kate Moore of the Geography Department at Leicester who drew all of the diagrams in exemplary style. Two anonymous referees did a lot to sort out both my ideas and my stumbling prose.

CHAPTER ONE

·MAPS AND MODELS·

INTRODUCTION

Everybody knows that geography is about maps, and every geography student knows that, among other things, it is now using quantitative methods of analysis to seek out and test hypotheses about how things vary spatially. Largely for historical reasons, the compilation and analysis of maps and the statistical analysis of what these maps portray have developed in isolation from each other, yet as one authority observes:

> To no small degree the recent quantitative analysis in geography represents a study in depth of the patterns of points, lines, areas and surfaces depicted on maps of some sort or defined by co-ordinates in two- or three-dimensional space. (Hagerstrand, 1973, p. 69)

For this reason, much recent quantitative analysis has become known as *spatial analysis* (Berry and Marble, 1968). The aim of this book is to present an elementary but integrated survey of this rapidly expanding field. It will be found to differ from many other quantitative geography and cartography textbooks in several important respects. First, because it is a geography book – not a statistics one – problems are examined by looking at the types of data used and the varieties of maps drawn, then at the analytical procedures that may be used to search for and isolate important spatial patterns. Second, no attempt is made to introduce techniques that treat data without reference to their spatial locations. This restriction has two undesirable but necessary consequences. It is assumed that the reader either has followed, or is following, a course in statistical analysis to the level of correlation and regression. (Where necessary, I have included references to basic statistical textbooks.) Also, although I have tried hard to avoid ego-massage by mystification, reading the completed text makes it apparent that in a few places the required mathematical level is not all that elementary. I do not worry overmuch at this: times have changed and, in the UK at least, many undergraduate geographers now have a background in advanced mathematics. As a teaching text, the book is probably most suited to a 10–20-session course in spatial analysis at the first- or second-year undergraduate level.

Third, no advice will be found on explicitly cartographic questions such as the choice of paper, pens or symbolism. These issues are covered in the relevant textbooks, for example by Monkhouse and Wilkinson (1971) and Robinson and Sale (1969). Fourth, the present book is in no sense intended as a contribution to philosophy, but in its concentration on techniques of quantitative analysis, I suppose that it may be labelled 'positivist'. To the phenomenologist or whatever it may prove old-fashioned and, perhaps, also uninteresting and irrelevant. I can only say that in research over the past fifteen years, I have not found such an approach to be either. As a profession, geography has never lacked bold, imaginative ideas. Quite the contrary: what usually is lacking is the technical skill necessary to sharpen up and test these ideas.

Finally, the book seriously tries to achieve coherence and consistency in its approach to map analysis. This is attempted using two fundamental organizing concepts discussion of which forms the greater part of Chapters 1 and 2. The first is that of *dimensionality,* particularly the dimension of length which enables us to classify map symbols. The second is that of *stochastic process theory.* Where possible, the adopted approach to map analysis is to compare observed measures of real-world map patterns with those expected under a null hypothesis of an independent random process. In effect, this is a comparison of an observed geography of interest, the map pattern, with the consequences of an hypothesis which asserts that no such pattern exists. In most practical work these tests will usually lead to a rejection of this null hypothesis in favour of an alternative which asserts that a pattern does exist. Such tests are, thus, very much a first step in any research work. Having decided that we have a pattern created by processes which are spatially dependent, the challenge to spatial analysts of the future is to isolate and model these processes.

THE NATURE OF MAPS

At first sight we believe we have a clear idea of what a map is, and what we mean by the process of mapping, but if we look at the variety of forms of representing data, it soon becomes evident that what initially seemed clear is actually remarkably vague. We seldom stop to examine what critical properties differentiate maps from diagrams, photographs, pictures or even language.

Characteristically, mathematicians have a precise, yet rather abstract, idea of what they mean by mapping, but it is an idea worth examining. *A mapping is a translation from one vector space to another,* and it can be illustrated as follows. Suppose that you purchase goods from a shop selling just three different sorts of article. Your requirements can be specified by three numbers telling how many of each article you want and, rather more abstractly, these numbers identify a unique point in a three-dimensional space, the first vector space. If you need two of article A, none of article B and five of article C, the purchase can be represented by the point (2, 0, 5) in a three-dimensional space. The bill is paid in a single quantity (money) so that, if the articles cost 10(A), 50(B) and 5(C) money units, the total is

$$(2 \times 10) + (0 \times 50) + (5 \times 5) = 45 \text{ units}$$

This can be represented as a point (45) along a line or one-dimensional vector space, and so we have traded from a one-dimensional money space into a three-dimensional article space specified by the three different types of purchase. A pawn-shop operates in a reverse direction, allowing multidimensional article space to be translated into one-dimensional money space. Notice that, whether buying or pawning, both require some rules, the *banking rules,* to connect the two spaces, and that the ordering of the items is significant. In the example the banking rules are the unit prices asked or paid for each article arranged in the order A, B, C. Furthermore, there is no reason why we should not extend our shop to stock 1000 articles (a 1000-dimensional space) and to accept payments not in one-dimensional money, but in a two-dimensional space of say, sheep and goats. Mathematically this is still a mapping of one space on to another, and we still require a banking rule to govern the operation.

The geographer concerned with spatial distributions also maps one vector space on to another, but prefers his maps to be capable of being drawn on sheets of paper, and is usually not much interested in sheep and goats. His first vector space is the real multidimensional world, his second the face of a sheet of paper on which the map is drawn. The requirement that maps are drawn on paper restricts them to a *three*-dimensional vector space, in which two dimensions are represented by the paper's length and breadth and the third by the lines and symbols drawn on it. The geographer further restricts his maps to a space representing the earth's surface, so that geographical mappings of phenomena are on to a space made up of two spatial dimensions that locate, and one quantity dimension that says what the phenomenon is, or how much there is of it, at that point. For example, a climatologist interested in the design of raingauge networks could draw a map of raingauge sites by placing a dot (indicating a raingauge) at all points where there are raingauges, and specify each point by its grid co-ordinates, so that a typical gauge would be recorded as (515, 925, 1), indicating one gauge at grid reference 515925. In other studies he could retain the grid co-ordinates and use the total recorded rainfall figure as the third variable, for example (515, 925, 1524), indicating a precipitation of 1524 mm at the same point.

In both cases our space has three dimensions, and each point can be represented by three dimensions (x, y, z), in which x and y are spatial co-ordinates and z is the phenomenon being mapped. We conclude that geographical maps are special cases of mathematical maps having three dimensions, two of which are used to locate phenomena. The trouble with this definition, which implies that just about anything of any size and located anywhere can be mapped by the geographer using any pair of locational co-ordinates, is that it is still too wide. In practice various conventions have been established which restrict even further what can be considered as a valid geographical map. First, geographers would now agree that the subject restricts its attention to a limited range of scale from the area of the earth's surface (5.101×10^{14} m^2) down to small areas of perhaps 100 m^2), so that the x and y co-ordinates are usually latitude and longitude on the earth's surface or some transformation of them. Second, the datum chosen is usually taken to be the plane of the earth's surface or a datum parallel to it, such as a fixed height in the atmosphere or depth in the soil. Third, the z-dimension, whether raingauges or questionnaire responses, is also restricted to a conventional range of topics thought to be of interest.

Maps are, therefore, representations of phenomena in a particular way and are subject to particular but largely unstated conventions. For them to be useful, it is necessary to know what they represent and over what area; this is equivalent to knowing the rules of a mathematical mapping. For the *x, y* co-ordinates, we need to know their linear *scale* and the *real-world area they represent.* This information is given by a scale bar and by base detail, such as latitude and longitude of the corners of the mapped area, a north point and, if the area is moderately large, the projection on to which it is drawn. For the *z* co-ordinate, we need a *key* that tells us what it represents and we take it on trust that the map itself is consistent with this. A map over which the symbolism were to keep changing its meaning would be worse than useless.

In his book *Theoretical Geography,* Bunge (1966) illustrates the importance of these various properties to our conventional notions of what constitutes a map. His 'method of traverses' starts with a clearcut example of what all would agree is a map and then exaggerates each property in turn until the map is abandoned in favour of what he calls a 'premap'. An example is a traverse in which the information content increases from a map to an aerial photograph, as in the sequence: (1) small-scale map (1 : 250,000); (2) large-scale map (1 : 50,000); (3) plan (1:2500); (4) aerial photomosaic with superimposed base detail; and, finally, (5) a single aerial photograph. At what point on this traverse does the map become a premap? An exercise in the Worksheet at the end of this chapter will illustrate the conventions adopted when we think of something as a map, and this is to assemble examples along a series of five such Bungian traverses, dealing in turn with scale, distortion, viewpoint, subject-matter and degree of abstraction and dimensionality.

WHY DO WE DRAW MAPS?

Maps are made and used for many purposes. Often we use them as an efficient data-storage medium in order to locate ourselves, other people, places and things, or to record administrative and other land boundaries; at other times we use them to demonstrate phenomena and relationships of special interest. To the geographer, most important of all is their use to suggest spatial relationships between phenomena across the same map, or from map to map, and to assist in the identification of the processes that produce spatial order and spatial differentiation.

In each case we rely on four major advantages that maps have over other forms of representation, although, as we shall see, each map use tends to stress one or other of these advantages more heavily than the others:

(1) Maps yield a synoptic or simultaneous expression of all the information represented.

(2) Because of this synoptic view, a map enables a number of important spatial properties that were not initially measured to be isolated. These include the orientation of phenomena, their shape and, most important of all, their relative location. By relative location, we mean where things are in relation to others, making up spatial patterns that cannot be seen unless a map is drawn. Maps, therefore, contain what has been called *spatial structure* (Unwin and Hepple, 1975). If spatial structure did not exist, it would be difficult to speak of 'geography', so that its detection and analysis is at the heart of the geographical sciences (King, 1969).

(3) A map that contained all the real things it mapped would be indistinguishable from reality itself, so as we move from the world of people on the earth to lines and symbols on paper, there must be a scale change, a selection of material and some generalization. Maps are, then, representations, or *models,* of the real world made in order to facilitate our understanding (Board, 1967). We can use them much as we use theories, to supply information, to predict and to analyse relationships in and about the real world.

(4) Finally, maps can be *communications devices* used to express and exchange ideas about the world. A bad map is like a bad book, giving its reader a distorted view of what its author intended, whereas a good map is like a good book with a clear message that is almost impossible to misinterpret.

WHAT MAPS CANNOT DO

These advantages of the map should not blind us to what maps cannot do. There are at least three important theoretical deficiencies in maps as models of the real world. First, they are essentially *static* and cannot be drawn to incorporate a time dimension, so that physical or human processes cannot be displayed directly. This creates a second difficulty in map analysis, that of structure/process asymmetry. When looking at a single map, we are examining one spatial structure that is the result of a spatial process. A *spatial process* is one which either has within itself, or which results in, spatial patterns and so will sort out height-dimension values, the z co-ordinates, into some pattern with respect to their x, y spatial co-ordinates. An example of a spatial process may be defined mathematically by

$$z_i = 2(x_i) + 2(y_i) \qquad (1.1)$$

in which x_i and y_i are the eastings and northings and z_i is the height of a point on the surface. As may be confirmed by drawing it for the range of x from 0 to 3 and y from 0 to 3, the map given by this process has a very regular spatial structure, and we can predict a value for z at all possible combinations of x and y. A second example of a spatial process is one controlled by one or more locational rules as in coalmine location; coalmines are sunk in areas underlain by coal so that any map of mines is certain to show a distinct structure, in this case a clustering of the mines on the coalfields.

There is a difficulty in the idea of a spatial process presented above that leads to a major problem in any spatial analysis which attempts to infer the process from the resultant structure. At the same time, ideas that are based upon this underlie most modern quantitative geography. The difficulty is as follows. The z-values forecast by equation (1.1) are uniquely determined by the mathematics, that is they result from a *deterministic process,* but geographic data are only seldom of this type. More often they appear to be the results of a *chance,* or *stochastic, process,* whose outcome is subject to chance variation which cannot be given precisely by a mathematical function. This chance element is inherent in processes involving the individual or collective results of human decisions, but even in physical geography a pattern that in theory is the deterministic result of a series of physical laws is often analysed as if it were a result of a stochastic process. Finally, the impossibility of exact measurement may introduce random errors into uniquely determined spatial structures.

Whatever the reason for this chance variation, the result is that the same process can generate many different spatial structures, as illustrated in Figure 1.1. The maps in (a) and (b) in the figure are contour maps based on the same underlying process as before

$$z_i = 2(x_i) + 2(y_i)$$

(see equation (1.1)), except that a random number in the range 0–9 was added to each point before the contour lines were drawn. For

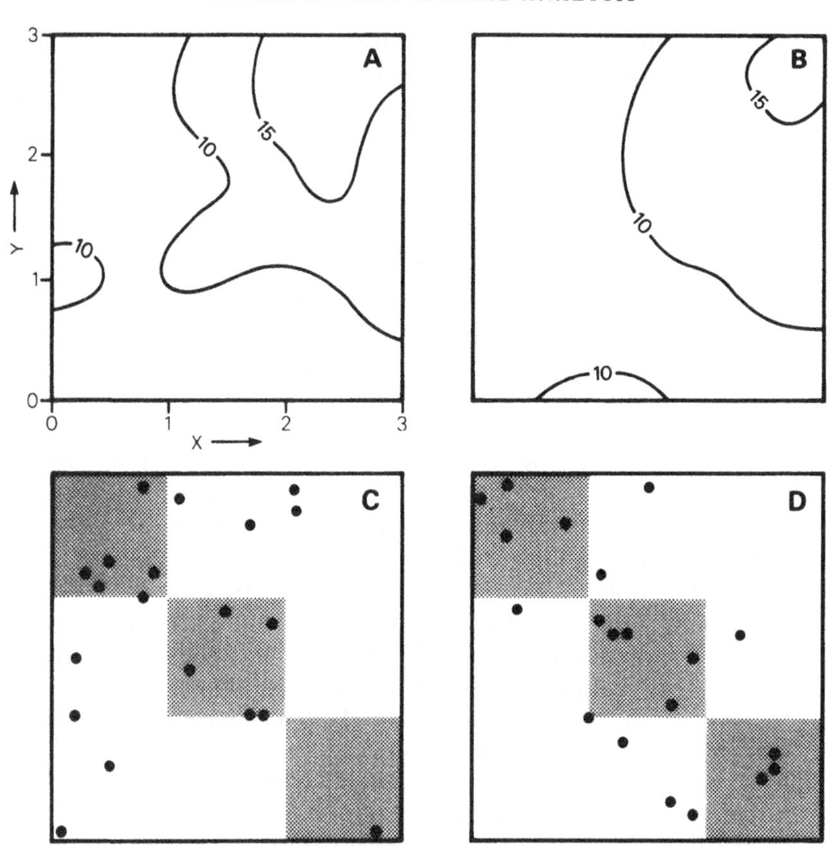

Figure 1.1 a, b, c and d Realizations of two stochastic processes.

example, the point at (1, 2) which deterministically had a z-value of 6 was disturbed by +1 in (a) and +2 in (b) in Figure 1.1 to give values of 7 and 8. The final maps are different but result from the same chance process. We call them *realizations* of that process. Despite their differences, both show the steady rise in z-values towards the north-west that was given by the deterministic process. Much the same point can be made about maps (c) and (d) in Figure 1.1 in which dots indicating the locations of some feature of interest have been drawn so that the shaded squares have three times the chance of receiving a point as those left unshaded. Apart from this rule, all the (x, y) values were chosen randomly, so the maps represent two realizations of the same process. In the real world, an example of this type of point process could be in the location of exploration oil rigs, where some areas are known to have a higher chance of striking oil than others.

A third deficiency in maps is that – because their makers are human – they can be used to lie. Maps can display misleading or inadequate data in equally misleading or inadequate ways, and the message they carry depends not only upon their compilers' preconceptions, but also on those of their users. Contrary to the often-expressed opinion that maps are somehow 'simple', one aim of this book is to show that they can be as complex and exciting as we choose to make them, forming the basic and rather sharp tools of all spatial analysis. As geographers, we spend our working lives using them, and from time to time it is inevitable that we shall get cut. Hopefully, the analytic methods to be described will help reduce the number of times this occurs.

RECOMMENDED READING

Berry, B. J. L. and Marble, D. F. (eds) (1968) *Spatial Analysis: A Reader in Statistical Geography*, Englewood Cliffs, NJ, Prentice-Hall.

Board, C. (1967) 'Maps as models', in Chorley, R. J. and Haggett, P. (eds) *Models in Geography*, London, Methuen, 671–725.

Bunge, W. (1966) *Theoretical Geography*, 2nd edn, Gleerup, Lund.

Hagerstrand, T. (1973) 'The domain of human geography', in Chorley, R. J. (ed.) *Directions in Geography*, London, Methuen, 67–87.

King, L. J. (1969) 'The analysis of spatial form and its relation to geographical research', *Annals, Assoc. Amer. Geogr.* 59, 573–95.

Monkhouse, F. J. and Wilkinson, H. R. (1971) *Maps and Diagrams*, 3rd edn, London, Methuen.

Muehrcke, P. C. (1978) *Map Use: Reading, Analysis, Interpretation*, Madison, JP Publications.

Robinson, A. H. (1952) *The Look of Maps*, Madison, University of Wisconsin Press.

Robinson, A. H. and Petchenik, B. B. (1976) *The Nature of Maps: Essays Towards Understanding Maps and Mapping*, Chicago, University of Chicago Press.

Robinson, A. H. and Sale, R. D. (1969) *Elements of Cartography*, 3rd edn, London, Wiley.

Sawyer, W. W. (1966) *A Path to Modern Mathematics*, Harmondsworth, Pelican Books, especially Chapter 3.

Thompson, M. M. (1979) *Maps for America*, Reston, Va, US Department of the Interior, Geological Survey.

Unwin, D. J. and Hepple, L. W. (1975) 'The statistical analysis of spatial series', *The Statistician* 23, 211–26.

Wright, J. K. (1942) 'Map makers are human: Comments on the subjective in maps', *Geographical Review* 32, 527–44.

WORKSHEET

Each chapter in this book has a Worksheet like the one below in which the reader is encouraged to examine, and, more importantly, use maps to illustrate general themes, or as a source of quantitative data. Each Worksheet suggests appropriate illustrative material chosen mainly from the maps I know best, those of the UK, but instructors using the book as a source text will doubtless be able to find their own examples. For the USA, an invaluable guide is M.M. Thompson (1979)

Continued on next page

Maps for America, Reston, Va, US Department of the Interior, Geological Survey.

Some Bungian traverses

The nature of the conventions adopted when we call something a map, can be illustrated by a series of Bungian traverses (see p. 5) from what are definitely maps to what are equally definitely 'premaps'. The following is a list of suggested materials to illustrate several such traverses:

Scale

Representations of the following:

(1) a region: OS 1 : 50,000, sheet 140, 'Leicester and Coventry';
(2) a city: OS 1 : 10,000, SK50SE;
(3) a small part of a city: OS 1 : 2,500 plan SK5904;
(4) a campus within a city;
(5) a room in a building;
(6) neutron diffraction patterns. Saibil, H. *et al.* (1976) 'Neutron diffraction studies of retinal rod outer segment membranes', *Nature* 262, 266–71.

A similar sequence using North American examples would be:

(1) Rand McNally *Road Atlas,* 'Alberta and British Columbia';
(2) 'Cranbrook-Lethbridge', 1:500,000, sheet 82SE, Department of Energy, Mines and Resources, Canada; and
(3) 'Lethbridge', 1:250,000, sheet 82H.

Distortion

(1) Large-scale map: J. C. Bartholomew (1975) *The World Atlas,* Edinburgh, Bartholomew, 29.
(2) Generalized map: 'Irish cultural influence in Europe', Dublin, Cultural Relations Committee of Ireland; or, more accessibly, Bunge (1966), figure 2.3.
(3) Cartogram: Skoda, L. and Robertson, J. C. (1972) ' "Isodemographic" map of Canada', *Geographical Papers* 5,

Continued on next page

Lands Directorate, Department of Environment, Ottawa, 34 pp. and map; Lockwood, A. (1969) *Diagrams*, London, Studio Vista, and New York, Watson Guptill, 100–1.

Having performed this traverse, now try to classify a map such as the *Hereford World Map* (Robinson and Sale, 1969).

Viewpoint

(1) Conventional orthogonal datum: J.C. Bartholomew (1966) *Plan of Central London*, Edinburgh, Bartholomew.
(2) Relief maps: Robinson, A. H. and Thrower, N. J. W. (1957) 'A new method of terrain representation', *Geographical Review* 47, 507–20; or the splendid town plans in *Staedte: 25 Bildplane von Hermann Bollman* (various dates) Braunschweig, Bollmann-Bildkarten-Verlag.
(3) Anaglyph map: ECU (1969) *Automatic cartography and Planning*, London, Architectural Press, 61.
(4) Perspective block diagram: ECU, op. cit., 63; or any output from the SYMVU computer program. A very attractive example of a perspective block diagram is that of Yosemite Valley reproduced on p. 235 of Thompson (1979).

Subject-matter and degree of abstraction

(1) Topographic map of area of earth's surface: J.C. Bartholomew (1975) *World Atlas*, Edinburgh, Bartholomew, 'Eurasia', 30.
(2) Map of a less concrete subject: isolines of height of 500 mb atmospheric pressure surface, *European Meteorological Bulletin, Deutscher Wetterdienst, Zentralamt, Offenbach am Main*, Daily, 3.
(3) Trend surface map: linear trend surface of corrie heights in Snowdonia, in Unwin, D. J. (1973) 'The distribution and orientation of corries in northern Snowdonia, Wales', *Transactions, Inst. Brit. Geogr.* 58, 85–97, figure 3(a).

Continued on next page

(4) Plot of a mathematical relationship: for an interesting comparison with exercise 3, plot the relation

$$z_i = 297.9 - 0.11(x_i) + 2.14(y_i)$$

over the rectangle defined by x from 0 to 90, and y from 0 to 180.

Dimensionality

(1) Relief model: 'Glasgow and Edinburgh', Oxford Plastic Relief Maps, Series 4, sheet 4, Oxford, Clarendon; 'Chambery' (1954) Carte Géologique Détaillée de la France, 1 : 50,000 type 1922, sheet 33.32, L'Institut Géographique National, Paris; or 'Princeton Quadrangle', 1:62,500 plastic relief map, Reston, Va, US Department of the Interior, Geological Survey.
(2) Topographic map: OS 1:50,000, sheet 115, 'Caernarvon and Bangor'; 'Ennis, Mont, Quadrangle', 1:62,500, Reston, Va, US Department of the Interior, Geological Survey.
(3) Dot map: Robinson, A. H. and Sale, R. D. (1969) figure 6.2.
(4) Bivariate statistical distribution: Haggett, P. (1965) *Locational analysis in human geography,* London, Arnold, figure 10.6(a).
(5) Features along a line: motorway strips in *Book of the Road,* London, Reader's Digest Association.
(6) Statistical frequency distribution: numerous examples in Gregory, S. (1978) *Statistical Methods and the Geographer,* 4th edn, London, Longman.

The influence of scale on generalization

Obtain the following sequence of Ordnance Survey maps (where possible) of the same area:

(a) 1 : 1250 plan
(b) 1 : 2500 plan
(c) 1 : 10,560 or 1 : 10,000
(d) 1 : 25,000
(e) 1 : 50,000 or 1 : 63,360
(f) 1 : 126,720
(g) 1 : 253,440
(h) 1 : 625,000

Continued on next page

Select a series of features such as houses, roads, railway stations and relief, and for each map record the method of representation used. Make a distinction between true-to-scale representations and those that use symbols. It is useful to organize the results as a matrix, in which rows represent features and columns the different map scales.

Thompson (1979:115–22) has such a matrix for maps of the USA together with a list of the available scales.

The map as a realization of a spatial process

This final exercise may seem tedious and unnecessary, but it is invaluable in that it very rapidly illustrates the idea of a stochastic process. First, draw the contours of the spatial structure defined by:

$$z_i = 3(x_i) + 2(y_i)$$

over the range of x from 0 to 3, and y from 0 to 3. Then use a coin-toss to decide whether to add (heads) or subtract (tails) 1 from the data points you used and contour the result. Finally, to add even more chance variability, repeat the experiment using random digits in the range 0–9. A suitable table of random numbers is table 8 of Lindley, D. V. and Miller, J. C. P. (1962) *Cambridge Elementary Tables,* Cambridge, Cambridge University Press.

CHAPTER TWO

·A TYPOLOGY OF MAPS·

INTRODUCTION

Chapter 1 examined the nature and role of maps without giving much attention to the type of information displayed, or the various kinds of map that can be drawn. The present chapter explores both these issues. First, it looks at the level, dimension and units of the available information, and second, it examines its spatial characteristics. Finally, based on these ideas, a typology that recognizes twelve fundamental types of map is developed.

LEVELS OF MEASUREMENT

The level of measurement of a spatially distributed variable is a basic control on the choice of map type, method of analysis and, ultimately, on the nature of the inferences that can be drawn from a study of that variable's spatial structure. Before presenting a list of the various levels of measurement that have been recognized, it is important to be perfectly clear what is meant by measure and measurement. When information is collected, *measurement* is the process of assigning a class or score to an observed phenomenon according to some set rules, but what is not always made clear is that this definition does not restrict us solely to assignments that involve numbers. It can also include the classification of phenomena into types or their ranking relative to one another on an assumed scale. You are reading a work which you assign to the general class of objects called books. You could rank it relative to other books on some assumed scale of merit as good, indifferent or bad, or even weigh it on a balance. In each case a measurement has been made, and it is apparent that this rather general view of measurement describes a process that goes on in our minds virtually all our waking lives as we sense, evaluate and store information about our environment. If this everyday process is to yield scientifically useful measurements, it is necessary to insist that they are made using an operationally *definable process,* giving *reproducible* outcomes that are as *valid* as possible. The first requirement implies that the measurer has some idea of what he is measuring and is able to perform the necessary operations; the second that repetition of the process yields the same results and gives similar results when differing data are used; the third implies that the measurements are true or accurate. If any of these requirements are not met, the resulting measurements will have only limited scientific use. In short, we need to know what we are measuring, there must be a predefined underlying scale on to which we can place phenomena, and we must use a consistent set of rules to control this placement.

Frequently in geography, we are interested in mapping concepts that are not readily measured and for which no agreed measurement rules exist. This is most common when the concept itself is very vague or encompasses a variety of possible interpretations. In physical work, we can readily measure altitude above sea level and slope, but it is much harder to measure the presence or absence of planation surfaces, because not all workers will agree that a given flat area is such a feature. In social work the concept of overpopulation cannot be measured simply by the population density, because it involves people's reactions, standard of living and the available resources. Work in behavioural geography and perception very clearly illustrates this kind of difficulty, because the concepts involved are in themselves vague (Downs, 1970). What, for example, do mental maps such as those produced by Gould and White (1974) really show? Do people evaluate where they might live, where they could live, or where they want to live?

The rules defining the assignment of a name or number to phenomena determine what is called the *level of measurement,* with different levels being associated with different rules. A particularly useful classification of measurement levels is that devised by S. S. Stevens (1946), who identified four basic levels: nominal, ordinal, interval and ratio.

Nominal

Because it makes no assumptions whatsoever about the values being assigned to the data, this is the lowest level in Stevens's scheme. Each value is simply a distinct category, serving solely to label or name the phenomenon. Thus we call certain types of animal 'sheep' and, as long as we remember the key, there is no loss of information if these are called 'category 2'. All that is required is that the categories are inclusive and mutually exclusive. By inclusive, we mean that it should be possible to assign all objects to some category or other ('sheep' or 'not sheep'), and by mutually exclusive, we require that no object be capable of being placed in more than one class. No assumption of ordering or of distance between categories is made, so that the numbers 1 (equivalent to the category bison), 2 (sheep) and 3 (other four-legged animals) do not imply that two bison are equal to one sheep. The numerical values 1, 2, 3 . . . ,*n* in nominal data serve merely as symbols and cannot be manipulated mathematically in any meaningful way. This obviously limits the operations, statistical and cartographic, that can be performed on them; but we shall see later that we can count them to form frequency distributions and, if they are spatially located, perform useful mathematical operations on their (x, y) locational co-ordinates. One of the simplest cases of nominal data is in dichotomies, such as yes/no, male/female, here/not here, that can be coded using the binary symbols 0 and 1.

Ordinal

In nominal measures there are no implied relationships between the classes. If it is possible to rank them consistently according to some criterion, then we have an ordinal level of measurement. An example is the classification of land into capability classes according to its agricultural potential. We know order, but not distance, along an assumed scale. Like nominal data, the usual mathematical operations cannot be performed, but some statistical manipulations that do not assume distances are possible. Ordinal scales have two important properties. First, it is essential that relationships between classes or objects are asymmetric. If A is greater than B, then B cannot also be greater than A. Second, the relationships are also transitive in that, if A is greater than B and B greater than C, A must also be greater than C.

Interval and Ratio

In addition to ordering, the interval level of measurement has the property that the distances between categories are defined, using fixed equal units. Thermometers employed to measure temperature normally measure on an interval scale, ensuring that the difference between, say, 25°C and 35°C is the same as that between 75.5°C and 85.5°C. This interval scale lacks an inherent zero and can, thus, be used only to measure differences, not absolute magnitudes. Ratio scales have in addition an inherent zero. A distance of 0 m really does mean no distance, unlike the interval scale 0°C, which does not indicate no temperature; and 6 m is twice as far as 3 m, whereas 100°C is not twice as hot as 50°C. This difference is best clarified by examining what happens if we calculate a ratio of two measurements. If place A is 10 km (6.2137 miles) from B and 20 km (12.4274 miles) from C, then the ratio of the distances is

$$\frac{\text{distance to B}}{\text{distance to C}} = \frac{10}{20} = \frac{6.2137}{12.4274} = \frac{1}{2}$$

irrespective of the units of distance used. Distance is fundamentally a ratio-scaled measurement. Interval scales do not preserve the ratios in the same way. If place B has a mean annual temperature of 10°C (50°F) and C of 20°C (68°F), we cannot claim that C is twice as hot as B because the ratio depends upon our units of measurement. In Celsius it is 20/10 = 2, but in Fahrenheit it is 68/50 = 1.36. Both interval and ratio data can be manipulated arithmetically and statistically in more or less the same ways, so it is usual to treat them together. Table 2.1 summarizes the differences between the four recognized measurement levels.

It should be noted that, although data may have been collected at one measurement level, it is often possible and convenient to convert them into a *lower* level for mapping and analysis. Interval and ratio data can be converted into an ordinal scale, such as high/low or hot/tepid/cold. What is generally not permitted is to collect data at one level and then attempt to map and analyse them as if they were at a higher level as, for example, by trying to add ordinal scores.

DIMENSIONS AND UNITS

Apart from the property of level, the measures also have the property of *dimensionality* and are related to some underlying *scale of units*. If we want to describe a stream, the variables we might consider important include its velocity, wet perimeter, cross-sectional area, discharge, water temperature, conductance, and so

Table 2.1 Levels of measurement

Level	*Basic operations*	*Examples*
Nominal	Determination of equality of class; counting items	Classification of objects, land-use categories
Ordinal	Determination of greater, less, or equality; counting items in a class	Land-capability classification, city ranking
Interval	Determination of equality or difference of interval; addition, subtraction	Temperature in deg C or deg F
Ratio	Determination of equality or difference of ratio; addition, subtraction, division	Distance, mass, precipitation

on. These measurable variables are some of its so-called *dimensions* of variablity. The choice of dimensions is a matter of the interests of the researcher, but in many problems in applied science it has been found that most can be reduced to combinations of three fundamental dimensions of mass, length and time indicated by M, L and T. A velocity dimension is a distance L divided by a time T taken to cover it, or LT^{-1}, and this is true irrespective of whether it is recorded in miles per hour or metres per second. Notice that LT^{-1} is simply a way of writing a length divided by a time L/T. Similarly, cross-sectional areas can be reduced to the product of two length dimensions, or L^2, discharge is a volume L^3 per unit of time T with dimensions L^3T^{-1}, and so on.

A very important class of variables are those which are non-dimensional and have values independent of the units involved. An angle measured in radians is the ratio of two lengths whose dimensions cancel out (LL^{-1}), to give no length dimensions and many of the well-known numbers in science (e.g. the Froude, Mach, Prandtl and Stokes) are also non-dimensional. Another important source of non-dimensional numbers is observations recorded as proportions of some fixed total. Non-dimensional variables are particularly useful in comparisons between sets of numbers, or in scaling model experiments.

Dimensional analysis is without doubt an extremely valuable method in any applied work. Because equations must be balanced dimensionally as well as numerically, the method can be used to check for the existence of variables that have not been taken into account, and even to help in suggesting the correct forms of functional relationships (see Douglas, 1969). To date, geographers have shown little interest in dimensional analysis, perhaps because in a great deal of human work no obvious fundamental dimensions have been recognized. Yet as Haynes (1975, 1978) has shown. there is nothing to stop the use of standard dimensions such as P (= number of people) or $ (= money), and their use can often suggest or clarify the correct forms of equations.

Finally, the measurements are related to a fixed scale of *units*, the standard scales employed to give numerical values to each dimension. In the past many systems of units have been used to describe the same dimensions as, for example, in distance measurement which has been made in Imperial units (inches, feet, miles), metric (metres, kilometres) and even the vaguely traditional (hands, rods, chains, nautical miles), giving a bewildering and confusing variety of fundamental and derived units. Although

Table 2.2 Features of SI units

Base units

There are seven base units:
the metre (m) for length;
the kilogramme (kg) for mass;
the second (s) for time;
the ampere (A) for electrical current;
the kelvin (K) for temperature;
the candela (cd) for luminous intensity;
the mole (mol) for amount of substance

Derived units

Formed by combining the base units. The unit of force can be produced by combining the first three base units. It is called the newton (N), equivalent to 1 mkgs^{-2}. Other approved names for derived units are the hertz (Hz), joule (J), watt (W), coulomb (C), volt (V), farad (F), ohm (Ω), weber (Wb), tesla (T), henry (H), lumen (lm) and lux (lx)

Multiples of common units

Decimal multiples of the SI units are formed using:

factor by which unit is multiplied	*prefix*	*symbol*
10^6	mega-	M
10^3	kilo-	k
10^{-3}	milli-	m
10^{-6}	micro-	μ

Note: For further details, see British Standards Institution, *The Use of SI Units*, PD 5686 (London).

many systems were used because of their relevance to everyday life and are often very convenient, in science they are unsatisfactory and dangerously confusing. It is recommended that use is made of the *Système International d' Unités* (SI) based on the familiar metric units, and summarized in Table 2.2.

THE GEOGRAPHICAL DIMENSION OF DISTANCE

It is clear that from the point of view of level, dimension and unit, there is virtually no limit to the data we can represent on a map; all that is needed is a z-value and a pair of locational co-ordinates (x, y) to tell us where the value occurs. There is a far more restricted choice of the dimension of this (x, y) locational information. As was outlined in Chapter 1 (pp. 3–5), maps are drawn on flat sheets of paper that have the fundamental dimensions of area L^2. At first sight, it might be concluded that maps can only display areas (L^2), but a moment's thought will indicate that in fact there is a choice of no dimensions and the length dimensions L, L^2 and L^3. Get a pencil and draw on a blank sheet of paper: every mark you make can be reduced to the operations of dotting a *point;* scribing a *line;* shading an *area;* and, if you are happy to accept a limited number of conventions, representing a volume by *surface.* We have the basic geographical building-blocks of points, lines, areas and surfaces, and the (x, y) values can be related to one or other of these. We can speak of point, line, area and surface data.

A point refers to a single place and can be conceptualized as having no dimensions of length, even though to be visible on a map, it must in practice have some area. The basic symbol used is the coloured dot. Lines are widely used and have the single dimension of length L, but like the dot need area in order to be visible. Line data specify features such as rivers, roads, boundaries, gradients and flows. Areas have two dimensions of length L^2 and involve the use of a shading, usually called a 'colour', to represent information collected over them. Finally it is possible to use dots, lines or areas to represent *surfaces* as in a contour map of altitude. The lines define a surface that encloses a volume of land giving us three dimensions of length, L^3.

The distinction made between area (L^2) and surface (L^3) data may not be immediately apparent and is one which is worthy of elaboration. Consider a map on which each county of England has been coloured in a graded series, with the intensity of the colour

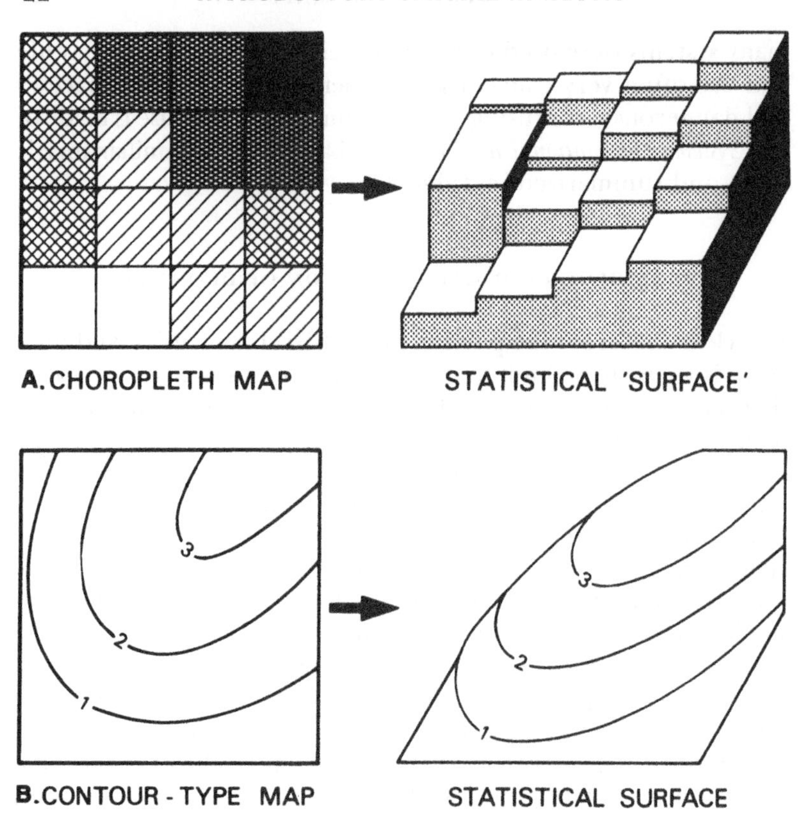

Figure 2.1 Translations between area and surface data:
a Choropleth map;
b Contour-type map.

related to some dimension of population. Are we dealing with area or surface data? Obviously, it is possible to conceptualize these data as the spiky statistical surface shown in Figure 2.1(a). Similarly, a contour map of an altitude surface can be regarded as a series of nested areas, all land 0–50, 50–100 m, and so on, as shown in (b) in the figure. Cartographically the distinction between surfaces and areas is somewhat arbitrary, but from a numerical point of view, there is an important difference between these data types that it is worth retaining. In *area data* the z-values relate to the density over some previously determined portion of the earth's surface, such as a parish, county, or country. The population of a county has no meaning unless we have also indicated the area to which it refers. In contrast, *surface data* refer to z-values that do not relate to some previously defined areal lattice, and where the data are continuous in space, a z-value existing for every possible (x, y) location as in climatic data like mean annual temperature.

TYPES OF DATA AND TYPES OF MAP

It will be recalled that, although we have an unlimited choice of units and dimensions, the data must fall into one of four levels of measurement and their (x, y) locational characteristics relate to one of four types of geographical data. Remembering that we have treated interval and ratio data as equivalent, a cross classification of these attributes gives us the twelve types of data exemplified in Table 2.3.

Thus, a city is a nominal assignment that occupies a zero-dimensional point location on a map, whereas the city population is a ratio measure that can also be assigned to a point. A road is a nominal measure that has the simple dimension of a length, whereas a major road would involve an ordinal assignment and a traffic count a ratio. The names of a series of counties give nominal area-based data. The same areas could equally be ordinally scaled into rich and poor or ratio scaled by their per capita income. Finally, nominal surface data would be generated where the values of the particular variable were present, in a continuous fashion at all (x,y) locations. Hence, the presence or absence (0/1) of precipitation or an assignment of the soil to one of a number of types gives us examples of nominal surface data with ordinal

Table 2.3 Types of data for mapping

	Dimension L			
	Point 0	*Line* 1	*Area* 2	*Surface* 3
Nominal	City	Road	Name of unit	Precipitation or soil type
Ordinal	Large city	Major road	Rich county	Heavy precipitation or good soil
Interval and ratio	Total population	Traffic flow	Per capita income	Precipitation in mm or cation exchange capacity

Note: Each cell gives one or more examples of the kind of data involved. The table is based loosely on Robinson and Sale (1969, figure 5.1).

Table 2.4 Types of map

Level	*Point* (*Chapter* 3)	*Line* (*Chapter* 4)	*Area* (*Chapter* 5)	*Surface* (*Chapter* 6)
Nominal	Dot map	Network map	Coloured area map	Freely coloured map
Ordinal	Symbol map	Ordered network map	Ordered coloured map	Ordered chorochromatic map
Interval and ratio	Graduated symbol map	Flow map	Choropleth map	Contour-type map

(heavy precipitation, good soil) and more obvious ratio (precipitation in millimetres, soil cation exchange capacity) equivalents.

It would be a happy conclusion to this chapter if it were possible to classify all maps into twelve types exactly corresponding to these twelve data types. Unfortunately, this is not strictly possible, because the map is not merely data ordered in a particular way, but also a communications device. From time to time, we must sacrifice theoretical elegance for visual effect. First, as we have seen, data collected at one level are often converted to a lower level before mapping so as to produce a clearer map. Second, dots and lines must both possess area in order to be visible. At first sight this seems a trivial matter, but at what point does a dot become an area and a line a volume? Finally, and most important, the cartographer must be careful always to maintain a clear distinction between the kind of data he is mapping and the symbolism he uses to represent these data. For graphic effect, data located at points need not necessarily be shown in all cases by point symbols, and area data may be shown by a stipple of dots, and so on. Continuous surfaces enclosing volumes are almost invariably shown by line symbols. Moreover, the symbols used often do double duty, representing for example by a point symbol both the quantity (ratio level), and the quality (nominal) of an object at a point.

Despite these qualifications, from the point of view of the spatial analyst rather than the cartographer, it is possible to recognize a typology of maps that roughly corresponds to the data types in Table 2.3. This is produced in Table 2.4, together with examples of each map type shown in Figure 2.2.

In concluding this chapter, notice that the cartographic distinc-

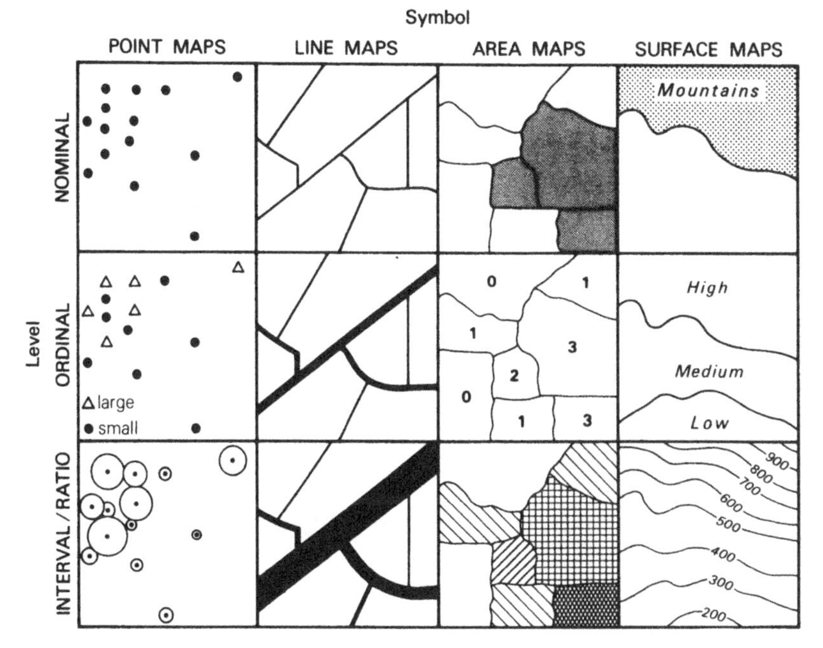

Figure 2.2 A typology of maps.

tion that is often made between abstract *thematic* maps and all-inclusive *topographic* maps disappears when we adopt the viewpoint of a spatial analyst. The topographic map can be seen as a composite of many different kinds of map overlain on the same base. In the distribution of point features such as churches and windmills, it has elements of point symbol maps, the rivers and transport networks exemplify network and ordered network line maps, while the land-use information and contour pattern are by our definition nominal and ratio surface maps, respectively.

RECOMMENDED READING

BSI (various dates) *The Use of SI Units,* London, British Standards Institution, PD 5686.

Douglas, J. F. (1969) *An Introduction to Dimensional Analysis for Engineers,* London, Pitman.

Downs, R. M. (1970) 'Geographic space perception', *Progress in Geography* 2, 67–108.

Gould, P. and White, R. (1974) *Mental Maps,* Harmondsworth, Pelican Books.

Haynes, R. M. (1975) 'Dimensional analysis: some applications in human geography', *Geographical Analysis* 7, 51–67.

Haynes, R. M. (1978) 'A note on dimensions and relationships in human geography', *Geographical Analysis* 10, 288–92.

Robinson, A. H. and Sale, R. D. (1969) *Elements of Cartography,* 3rd edn, London, Wiley.

Stevens, S. S. (1946) 'On the theory of scales of measurement', *Science* 103, 677–80.

Taylor, P. J. (1977) *Quantitative Methods in Geography,* Boston, Houghton Mifflin.

WORKSHEET

(1) Identify an appropriate level of measurement for each of the following properties of interest to the geographer: church, woodland, stream flow, elevation, slope, position of stream within drainage network, journey to work.

(2) Table 2.3 displayed data types of different levels of measurement, beginning with the nominal category exemplified by 'city', 'road', 'name of unit' and 'precipitation'. Attempt to fill out a similar table, beginning with 'factory', 'stream', 'Labour constituency' and 'freezing'. What cartographic symbolism would you use to map each of these?

(3) As outlined on pp. 24–5 topographic maps are sometimes assumed to differ from thematic maps but equally can be regarded as a series of superimposed thematic maps on a common base. An interesting exercise is to count the number of distinct thematic maps that can be

Continued on page 28

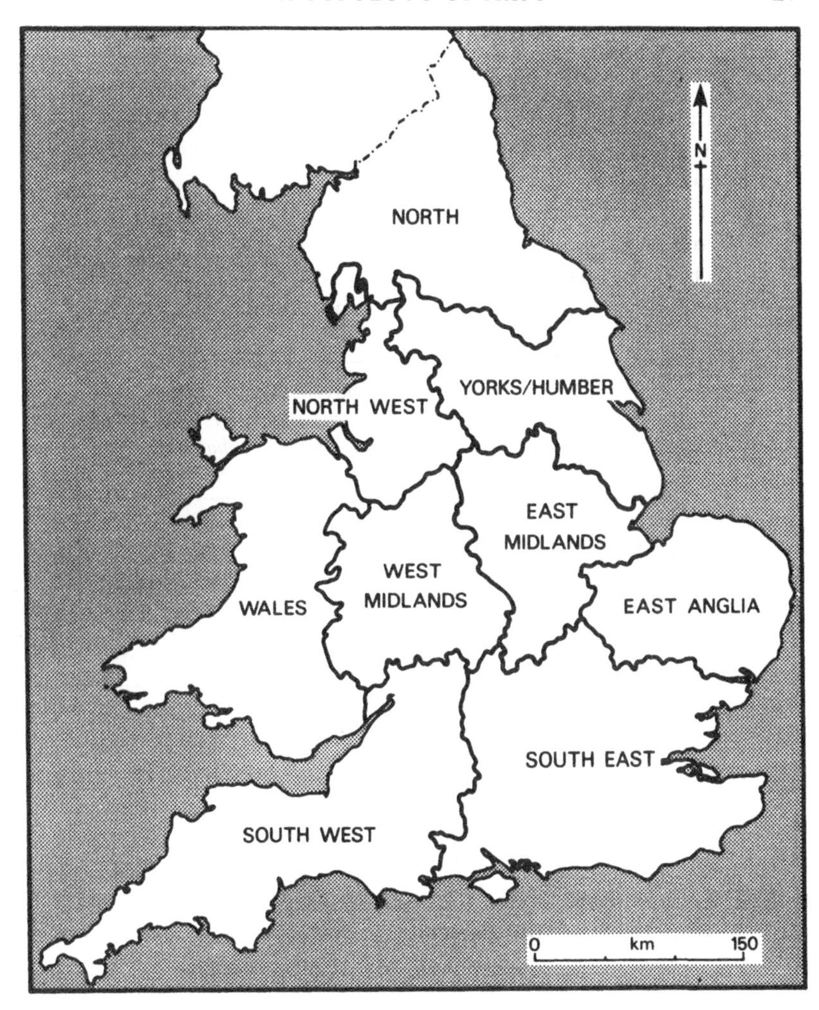

Figure 2.3 The standard regions of England and Wales.

derived from an Ordnance Survey 1 : 50,000 sheet or a US Geological Survey 1:62,500 sheet. (Different workers are unlikely to agree on this!)

(4) The data below are for the standard regions of England and Wales and are exclusively interval or ratio scaled. By choosing appropriate break points, reduce each variable to an ordinal and then a dichotomous nominal scale:

Region	*Number of days per year of illness per male at risk*	*Average weekly household income 1974–5 (£)*
North	25.3	60.8
Yorkshire and Humberside	21.5	59.5
East Midlands	15.9	64.0
East Anglia	12.9	63.3
South East		
Greater London	10.5	72.2
Other South East		75.9
South West	16.0	63.4
West Midlands	14.8	66.7
North West	21.0	61.3
Wales	32.2	60.3

(5) Using copies of the outline provided in Figure 2.3, map your results from exercise 4.

CHAPTER THREE

·POINTS ON MAPS·

INTRODUCTION

The simplest map we can draw has on it a number of dots, one at each point where a specified object is located. An example (showing the distribution of churches in south Leicestershire) is given in Figure 3.1. In the terminology of Chapter 2, each dot represents a nominally scaled value at a point location. Habit leads us to emphasize this positive side of things, but it should also be noted that the absence of a dot tells us where the object is absent. Because small round dots are most often used these maps are called *dot maps,* but there is nothing to stop the cartographer from using any other reasonably compact symbol – triangles, squares, small drawings of the object, and so on – the basic operation is still that of 'dotting'. If we allow the size of the chosen symbol to vary systematically, it is possible to show data measured at the ordinal, interval and ratio scales, giving a *proportional symbol map.* Collectively dot maps and proportional symbol maps are referred to as *point symbol maps.* Initially, both might seem to give an excellent visual impression of spatial pattern, but their apparently easy compilation, simplicity and clarity should counsel caution and there are some very important practical and theoretical issues involved in their use. Practically, it is necessary to examine various cartographic problems of dot size and value; theoretically, there is a need to examine what each point symbol represents, the method by which it is located, and how to use the evidence of the completed map to infer the spatial process that created it.

THEORY AND PRACTICE IN POINT SYMBOL MAPPING

The theory of dotting is straightforward. All we need do is place the chosen symbol at each (x, y) location of the object whose distribution is of interest. The only cartographic considerations are

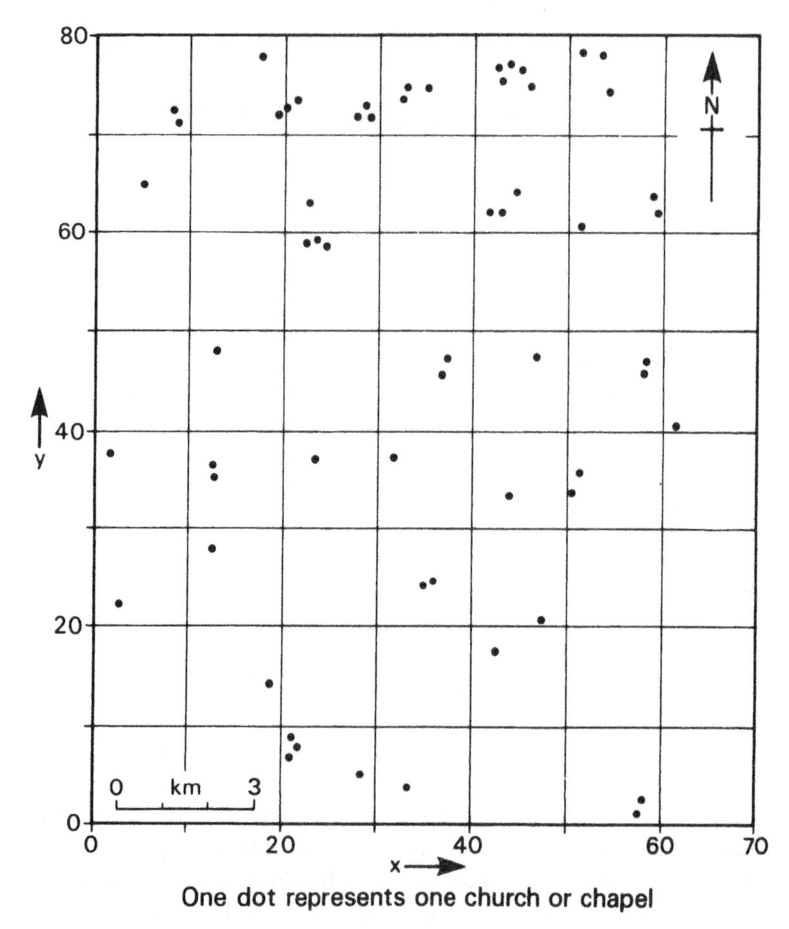

Figure 3.1 A dot map: distribution of churches in south Leicestershire.

the nature, size and colour of the symbol used. Because there is only one symbol for each and every earth object this is a 'one-to-one' mapping, the number of symbols being the same as the number of objects represented. The simplest symbol we can imagine is a small circular black dot centred on the location involved and with a size chosen according to cartographic criteria. The dots should be large enough to be visible on the completed map, even under any intended reduction, but equally should not be so large as to cause adjacent symbols to coalesce. It is a good idea to use an absolutely standard size, as given by a plastic template or by sheets of uniform adhesive symbols. The final visual impression depends upon the relative areas of black and white. Ideally, as more and more dots are placed in an area, the denser should seem the result in exact linear proportion, but unfortunately experiment shows that we do not perceive density in this way. In his classic article on 'Dotting the dot map', J. R. Mackay (1949) suggests that with a wide dot spacing there is a rapid change in apparent density as more dots are added, but as the dots get closer together the perceived change is less until the point at which coalescence occurs, which is again associated with a large perceptual change.

In contrast, dot location would seem a simple matter because each dot has its centre at a dimensionless point, representing a point object. In practice, only rarely will the objects represented occur at points. Usually they will have an area that can be ignored, if a sufficiently small fraction of the total mapped area, but if this area is significant, as for example large factories or mineral deposits, care should be taken to locate the dots sensibly.

Thus far the argument has assumed that the cartographer has no control over the number of dots, this being simply the number of earth objects to be represented. However tradition, and the requirement of visualization, have led to the use of dots to represent distributions that do not allow such simple one-to-one procedures. If we have a list of data collected over a specific area such as a total population or acreage under a crop, then it is possible to allow each dot to represent a previously defined quantity or number of objects, thus giving a 'many-to-one mapping'. The result is a *dot density map,* such as Figure 3.2 which shows the 1971 distribution of population in Leicester. The original data were the populations of 130 grid squares each of 1 km^2, and these have been mapped using a dot distribution value of one dot for each 100 persons. The map seems to give a good visual impression of population density variation across Leicester.

One advantage of the dot density technique is the control it gives over the overall visual impression. Changes in the distribution value will inevitably change the number of dots used, but unfortunately there are no obvious rules on how to choose an appropriate value. Some authors suggest that, for given dot size, a distribution value should be chosen that will cause the dots just to coalesce in the most dense areas. Others argue that the original data should be recoverable from the map, and that this implies that each dot be distinct and countable. A third consideration is that of convenience and speed of production. Too small a dot size and

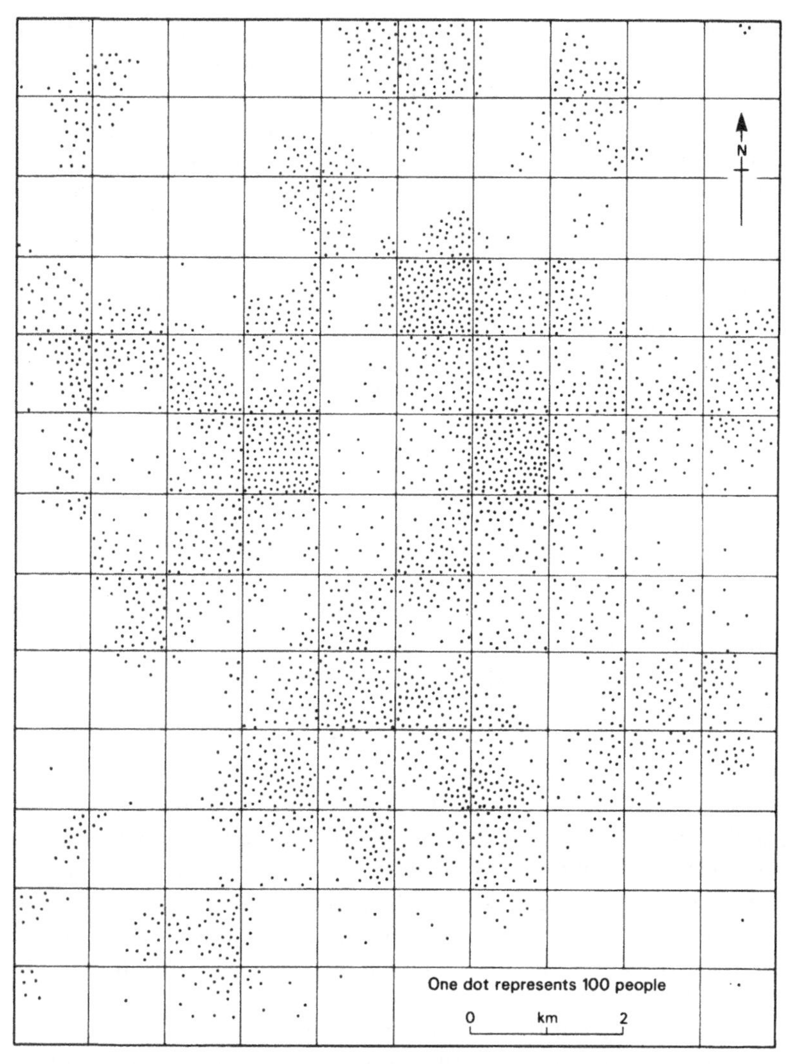

Figure 3.2 A dot density map: distribution of population in Leicester, 1971.

value gives an accurate map which is laborious to produce and that conveys a poor impression. Too large a size and value gives an easily drawn map that overgeneralizes the data. One possible solution to this problem is to fix the total number of dots in advance, defining each to represent some proportion of the total and leading to a *percentage dot density map* (Mackay, 1953). The major disadvantage of all dot density maps is that within any of the subareas for which the data were obtained, the exact location of any dots is arbitrary. In the absence of any other information about the underlying distribution, the cartographer has no choice but to apply the dots evenly over the area. When other information is available, the dots may be located 'sensibly' to reflect this. Figure 3.2 was drawn this second way, using as a guide the distribution of housing shown on a 1 : 50,000 topographic sheet of the same area. From a statistical viewpoint, a major consequence of this freedom to choose dot number and location is that most of the methods of analysis commonly used which assume accurate distances between dots are not applicable to this type of map.

Whether the dot map is a result of one-to-one or many-to-one mapping, it can be seen that it gives a useful picture of spatial distributions, but such maps are limited in that the data are either strictly nominal (like the churches in Figure 3.1), or interval and ratio regarded as nominal (each 100 persons in Figure 3.2). Often we may wish to represent point-located interval or ratio scaled data for which single dots cannot be used. Examples might be the outputs of a series of factories, the numbers of people they employ, or even the populations of cities over a wide area. Notice that in each case the data are intrinsically point-valued and are not samples drawn from an underlying continuous distribution. The method adopted to show these types of data uses *proportional point symbols*, the size of the symbol used being varied systematically according to the value of the variable to be represented. The dot maps discussed previously are, thus, a special case in which each symbol has a value of one of this more general type.

The most commonly employed method uses the circle and systematically varies its area according to some function of the values to be represented. The arithmetic involved is quite simple. The desired area of the circle is to be proportional to some value, that is: area of circle $\propto$ value. If we let l_r be the required radius of the circle and z the value, then since the area of a circle is πl_r^2, we have $\pi l_r^2 \propto z$. Hence the required radius will be proportional to the square root of the z-value, according to some convenient scaling factor or constant of proportionality k:

$$l_r = kz^{0.5}$$

For those unfamiliar with this method of indicating a square root, all we have done is to raise z to the power of one-half, 0.5, shown as

$z^{0.5}$, which is exactly the same as a square root; a cube root would be $z^{0.33}$, and so on.

Proportional point symbols using this square root relation are a very effective way of illustrating point-located interval and ratio-scaled data. A complex example is given in Figure 3.3, which shows world production and production capacity of uranium. However, like its simpler relative the dot map, this apparent cartographic clarity obscures some rather troublesome technical details associated with the constant of proportionality, the symbolism used, the overlapping of symbols and the nature of the key. In essence a proportional symbol map is expected to be able to allow its user to estimate the actual numerical data values from the evidence of the map symbols. This involves many problems, the simplest of which concerns choice of the constant of proportionality. The value chosen for k will depend on the units used for z and l_r and controls the range of circle sizes involved. It is, therefore, a very important control over the final product. Usually a suitable value is chosen by trial and error, so that the range of sizes is neither too large (giving some very large, possibly overlapping symbols), nor too small (giving symbols which tend to appear the same size).

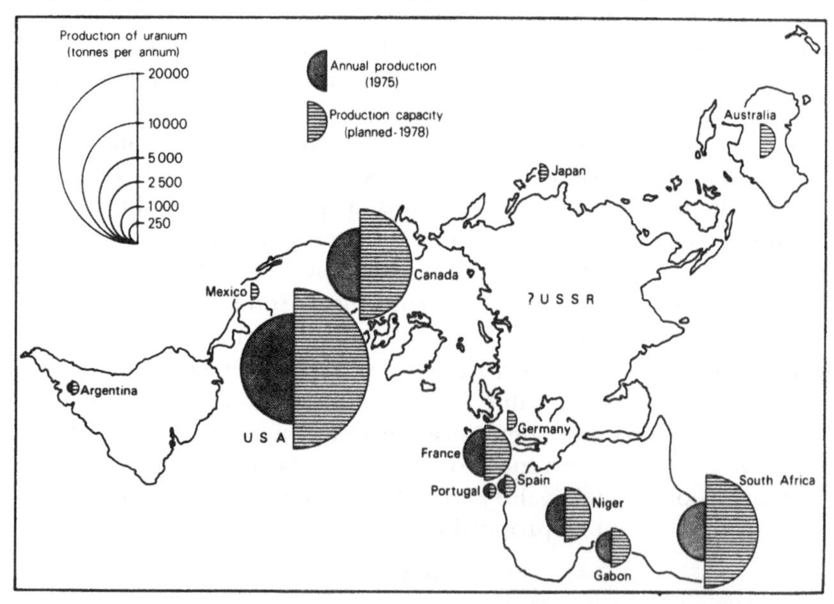

Figure 3.3 A proportional symbol map: world uranium production and capacity. (By courtesy of P. R. Mounfield.)

The choice of functional form is not so straightforward. We have *assumed* that a circle with square root relation is appropriate; it is convenient to find on a calculator and circles are easily drawn and the most compact of all plane figures. The difficulty with this method is, since our brain and eyes do not perceive increases in circular area as a linear function, that recovery of the original mapped data is difficult. Most people underestimate the sizes of the larger circles relative to the small ones, so that the square root method tends to devalue the larger data. Several solutions to this problem have been suggested:

(1) The square root relation is but one of an infinite number of possible power functions that we could have used. In general, we can write $l_r = kz^b$ in which, like k, b becomes a constant that we are free to choose to suit our purposes. To correct the perceptual imbalance of underestimating larger sizes, Robinson and Sale (1969) suggest the use of $b = 0.57$. This value is not particularly sacred, other workers (Williams, 1977) have proposed a value of 0.7.

(2) A second possibility is to change the symbol used to squares, triangles, etc., or even to attempt to reproduce three-dimensional spheres of volume proportionate to z, that is with $b = 0.333$. In all cases the compactness of the circles is lost and it is likely that even more severe perceptual problems will be involved (see for example Williams, 1956). They are also harder to draw!

(3) Mixing symbol types on the same map is permissible, as for example a mixed dot and proportional circle map, using dots in data-sparse areas, circles in data-dense zones.

(4) In his book *Graphical Rational Patterns,* Bachi (1968) introduces ingenious symbolism that seems to circumvent many of the problems we have noted above. This uses a unit square within which a conventional arrangement of symbols indicates the value involved – in a manner similar to the markings on dominoes. A variety of symbolism has been developed, and the interested reader is referred either to Bachi's book, or to the more accessible example in Lewis (1977, figure 2.2).

(5) A final possibility is to improve the key so as to show a good range of representative circles, as suggested by Dobson (1974) and Cox (1976).

Whatever the symbolism used, each symbol will inevitably occupy a distinct area on the map surface, so that neighbouring symbols may overlap to a greater or lesser extent. In principle this can be indicated by preserving each symbol outline, but in practice such overlaps are not easy to draw and can be avoided by a suitable choice of k.

The final point symbol map to be considered here still makes use of located proportional symbols but segments each symbol to give a *pie chart map*. These maps are used to show data that make up proportions of a whole, as for example the proportions of different products in the total output of a factory. Each symbol is made proportional to the total as before, but is then subdivided into two or more differently shaded parts one for each of the components. Almost invariably sectors of a circle are used, hence the name '*pie chart*' given to each symbol. These are very compact and effective, but to the general public they can be difficult to understand and are certainly open to all sorts of perceptual problems even on the part of experienced map users.

In summary, it can be seen that a great many point symbol maps can be drawn, ranging from simple dot maps, through dot density and percentage dot density maps to proportional point symbols, graphical-rational patterns and pie chart maps. In the next section, some ways are introduced by which the maximum amount of information can be extracted from these maps.

SOME DESCRIPTORS OF PATTERN

In the past a point symbol map would often have been the end-product of an investigation. This may have been a result of the labour involved in the drawing, but it also reflected a traditional descriptive approach in geography. Although it would be foolish to reject in its entirety this subjective approach, it did lack both precision and objectivity. A major danger was that the eye saw pattern where more detailed analysis would have shown that none existed, or that extant patterns that did not fit the investigator's preconceived notions would remain undiscovered.

Nowadays, a map is usually the starting-point of an analysis. The next stages are to summarize and describe the distribution, then to examine whether or not a recognizable pattern exists. Our dots and circles constitute a point pattern, so this, and the following sections, are concerned with methods of point pattern analysis. The techniques to be outlined will apply to any true one-to-one dot map, irrespective of what is displayed and, within limits, of scale also. Initially, we shall consider some numeric measures that can be used to describe point symbol maps, postponing any statistical considerations until later (see pp. 48–60).

Dot maps are often thought to be lacking in precise quantitative information, but this is not strictly correct. Although the nominally scaled dot restricts our information about what is being mapped, we do possess precise information about each object's (x, y) location and our techniques can exploit this by examining their spatial pattern. In attempting to understand point pattern analysis, it is essential to fully grasp the concept of *pattern*. Pattern is that characteristic of a spatial arrangement given by the spacing of individuals in relation to one another. It is, therefore, not related to the shape of the study area in which it is found and can be contrasted with the property of *dispersion*, which is the spacing of objects in relation to an enclosing shape. The *density* is the property of dispersal relative to an area but is independent of the area shape or the dispersion of the objects within it. Figure 3.4 attempts to indicate these differences: two identical patterns are shown in (a) with differing density, in which the shape of the containing area and hence the dispersion is not considered; (b) shows identical patterns and dispersions relative to a square area but with an obvious difference in density; while (c) shows the same six-point density and pattern but very different dispersions.

Our problem is to characterize these properties of density, dispersion and pattern, using objective repeatable measures which can be used to compare map with map and point pattern with point pattern. In general three distinct approaches seem to have been taken to solve this problem, one based on density considerations, another on distance measures and a third on interpoint distance and direction.

Measures based on density

The simplest measure we can adopt is the crude density or number of points per unit of area:

$$\text{density} = d = \frac{\text{number of points}}{\text{area containing them}} = \frac{n}{a}$$

The fundamental dimensions of a density are, thus, L^{-2}. An

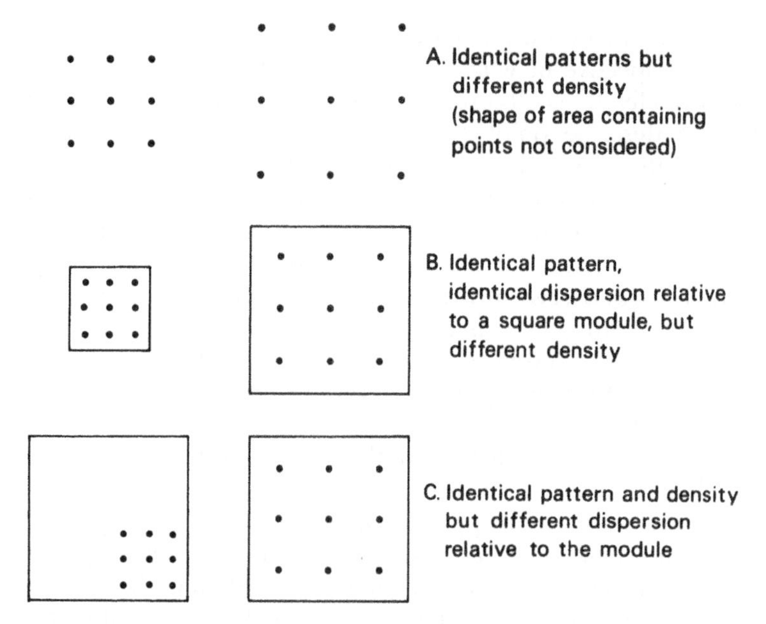

Figure 3.4 a, b and c Pattern, density and dispersion.

alternative that is sometimes used is the reciprocal $1/d = a/n$, which has dimensions L^2 and is the average area occupied by each point. Figure 3.1 has sixty points in an area of 56 km^2, giving a crude density $d = 60/56 = 1.07143$ km^{-2} and an average area per point of $56/60 = 0.933$ km^2. Inspection of Figure 3.4 makes it obvious that the values of these measures could have originated from an infinite variety of patterns and that they are very dependent on the area over which the calculation is made. They tell us very little about dispersion and pattern.

An alternative is to count the number of objects falling into a series of identically shaped and equal-size subareas – or *quadrats* – laid across the map. In effect, this produces a picture of density variation across the area, but the statistical frequency distribution of the number of quadrats containing $m = 0, 1, \ldots, n$ points is a very useful summary of the pattern, and forms the basis of a number of statistical tests, described on pp. 55–8. The process is illustrated in Figure 3.5, which shows a dot distribution over which a grid of quadrats has been laid and the counts summarized as a frequency distribution. It is evident that the shape of the frequency distribution will reflect some of the properties of the original

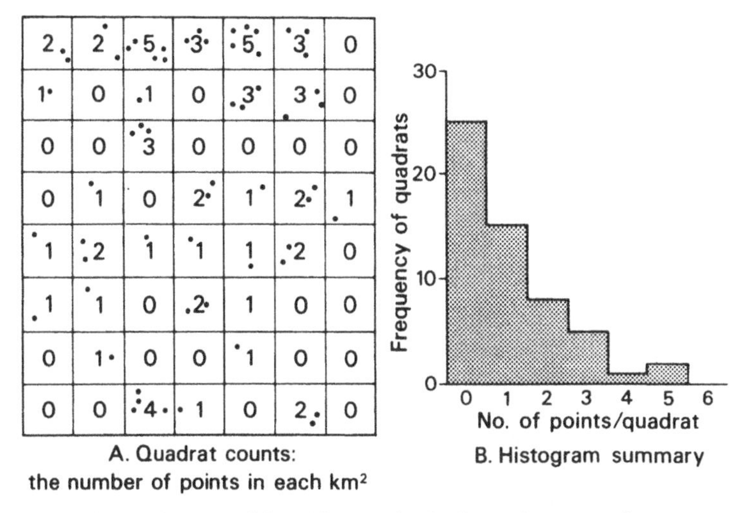

Figure 3.5 a and b The method of quadrat counting.

pattern, so conventional descriptive statistics such as the mean and variance can be used to compare patterns. A particularly useful measure is the ratio of the variance to the mean of the distribution, as calculated in Table 3.1 at 1.5903. Had the dots been completely uniformly spread with an equal number in each cell, the mean would be unchanged, for it depends only on the number of points, n, and the number of cells, k; but the variance about this would have been zero, giving a variance/mean ratio of zero. Clustered patterns of the same crude density will tend to increase the spread, and hence the variance, to give a high variance/mean ratio.

Quadrat methods have been extensively used both in plant ecology (Greig-Smith, 1964), and in geography (see Rogers, 1974; Thomas, 1977, for reviews), but in use a number of difficult problems have been isolated. The resulting frequency array must strongly depend on the quadrat size, shape and origin. As quadrat size is reduced the chance of any one containing a point is similarly reduced, so the question of the most appropriate size is an important one. One solution is to repeat the counts, and any summary statistics based on them, for a variety of quadrat sizes, as for example in Greig-Smith (1952), Rogers (1974) and Mead (1974). The question of quadrat shape has been little considered by geographers, a square lattice being almost universally used, but in principle there is no reason why other shapes should not be used, the only constraint being that shapes should be capable of packing

Table 3.1 Calculation of mean and variance for the histogram in Figure 3.5

Number in class (m_i)	*Frequency of cells* (f_i)	$(f_i m_i)$	*Deviation from mean* $(m_i - \overline{m})$	$(f_i(m_i - \overline{m})^2)$
0	25	0	−1.071	28.699
1	15	15	−0.071	0.077
2	8	16	0.929	6.898
3	5	15	1.929	18.597
4	1	4	2.929	8.577
5	2	10	3.929	30.867
	56	60		93.714

$$\text{Mean, } \overline{m} = \frac{\sum f_i m_i}{\sum f_i} = \frac{\text{no. of points}}{\text{no. of cells}} = \frac{60}{56} = 1.071$$

$$\text{Variance, } s^2 = \left[\sum f_i(m_i - \overline{m})^2\right] \Big/ \left(\sum f_i - 1\right) = \frac{93.714}{55} = 1.704$$

$$\text{Variance/mean ratio} = \frac{s^2}{\overline{m}} = \frac{1.704}{1.071} = 1.5903$$

completely to cover the study area. A final problem, particularly if statistical analysis is intended, concerns two quite different ways of assembling the frequency distribution. In geography the most commonly used is the one we have described, in which the area is exhaustively partitioned into a grid of contiguous quadrats. Every point will be counted just once, so the method can be called *quadrat censusing*. An alternative method often used by ecologists is to select a single quadrat and to place this randomly in the area k times, counting the frequency at each step to give the frequency distribution. This is *quadrat sampling*. Readers who wish to follow the ecological literature should be careful to keep this distinction in mind. Those who do not should recognize that much of the statistical theory of quadrat analysis is derived from a sampling, not a censusing, model.

Measures based on distances

A second type of pattern description uses a spacing approach based on the distances between the point symbols. *Distance* is one of the fundamental spatial concepts and like many such concepts, at first sight, seems easy to understand and measure; but on closer inspection, turns out to be much more difficult. Measuring the straight-line distance between two points presents few problems, if Euclidean geometry is used. The fundamental characteristic of a Euclidean scheme is that the well-known Pythagoras theorem holds good in any right-angled triangle, as illustrated in Figure 3.6(a). In the triangle it is easy to show that

$$l_c^2 = l_a^2 + l_b^2$$

It follows that the straight-line distance between the points labelled 1 and 2 can be found from their spatial co-ordinates. As the figure shows, side l_a has a length $(y_1 - y_2)$, side l_b is $(x_1 - x_2)$ and l_c can be found from

$$l_c = [(x_1 - x_2)^2 + (y_1 - y_2)^2]^{0.5}$$

Notice that this distance is only exact subject to two important assumptions. First, we have assumed that a Euclidean geometry is appropriate, and this is the case for problems in practical surveying in a small area. For large areas we should have to consider distances along Great Circles, and for problems involving human behaviour, better theoretical understanding might be obtained by restating distance as cost, time, or even perceived distance, which are unlikely to be Euclidean. Second, we have only found the straight-line, 'crow-flying' distance; again this might not be appropriate to the particular research problem. We might, for example, be more interested in road distance, distance down a river, and so on. Measuring distance along irregular curves like a coastline is a classic problem in mathematics, the difficulty or, as it is called, *Steinhaus paradox*, being that the more accurately we measure, the longer the line appears to get. (Exercise 2 in the Worksheet at the end of the chapter explores this paradox further.)

The mean centre. The first distance-based measurement that can be applied to a dot point symbol map measures the central tendency of the pattern and is called the mean centre. It is found simply by calculating the arithmetic means $\bar{x}$ and $\bar{y}$ of the spatial co-ordinates of the n points. In symbols, the mean centre is located at

$$\left(\sum x_i/n, \sum y_i/n\right) = (\bar{x}, \bar{y})$$

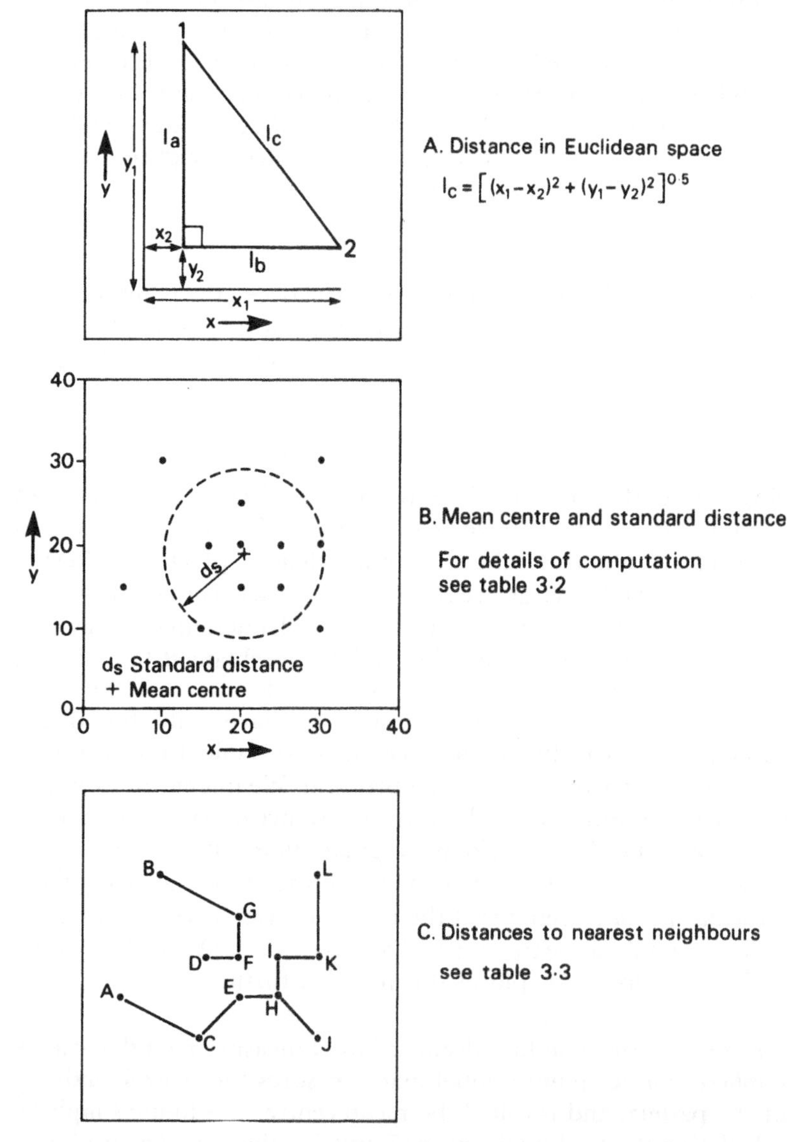

Figure 3.6 a, b and c Some distance measures of pattern.

An example of a point pattern with its mean centre is shown in Figure 3.6(b), and the relevant arithmetic in Table 3.2. Mean centres are useful summaries of a point pattern, particularly when the cartographer wishes to record the changing pattern of a single distribution over time, or to summarize differences in the distribution of a series of nominally scaled point data.

This simple notion can be extended to cover interval or ratio scaled point data represented by a proportional symbol map. The solution, in this case, is to weight each point by the quantity z_i located there:

$$\bar{x} = \Sigma z_i x_i / \Sigma z_i \text{ and } \bar{y} = \Sigma z_i y_i / \Sigma z_i$$

giving a *weighted mean centre.*

The standard distance. Just as in ordinary statistics we measure the scatter of observations on a single variable about their mean by the standard deviation, so we can measure the spatial dispersion of a point pattern by its standard distance d_s. There are two ways of calculating this. The first is to find the square root of the average square of the distances from every point to the mean centre, that is:

$$d_s = \left[\sum l_{ic}^2 / n\right]^{0.5}$$

where l_{ic} = distance of the *i*th point from the mean centre. It is apparent that l_{ic} may be found using the Pythagoras theorem as:

$$l_{ic} = [(x_i - \bar{x})^2 = (y_i - \bar{y})^2]^{0.5}$$

The calculations for the point pattern shown in Figure 3.6(b) are laid out in the rightmost columns of Table 3.2. Even using a modern hand calculator, this is a fairly tedious, error-prone business. Notice that the final column of distances (l_{ic}) is redundant. Strictly speaking, it is only necessary to find the sum of the $(x - \bar{x})^2$ and $(y - \bar{y})^2$ columns in order to find the Σl_{ic}^2 we require, but it does form a useful computational check. The calculated d_s is 10.019 length units; this can be drawn as a circle of this radius centred on the mean centre, as shown on Figure 3.6(b).

An alternative method of calculation that yields a little more information is shown at the bottom of Table 3.2. Here we have calculated the variance along the x and y axes independently and pooled the two results to give the overall standard distance. Although this overall d_s is the same as that calculated by the first

Table 3.2 Calculation of mean centre and standard distance

Point	x	y	$(x - \bar{x})$	$(y - \bar{y})^2$	$(x - \bar{x})^2$	$(y - \bar{y})^2$	l_{ic}
1	5	15	−15.5	−4.166	240.25	17.36	16.05
2	10	30	−10.5	10.833	110.25	117.36	15.09
3	15	10	−5.5	−9.166	30.25	84.03	10.69
4	16	20	−4.5	0.833	20.25	0.69	4.58
5	20	15	−0.5	−4.166	0.25	17.36	4.20
6	20	20	−0.5	0.833	0.25	0.69	0.97
7	20	25	−0.5	5.833	0.25	34.03	5.85
8	25	15	4.5	−4.166	20.25	17.36	6.13
9	25	20	4.5	0.833	20.25	0.69	4.58
10	30	10	9.5	−9.166	90.25	84.03	13.20
11	30	20	9.5	0.833	90.25	0.69	9.54
12	30	30	9.5	10.833	90.25	117.36	14.41
	246	230	0.0	0.000	713.00	491.66	

$$\left.\begin{aligned} \bar{x} &= \frac{246}{12} = 20.500 \\ \bar{y} &= \frac{230}{12} = 19.166 \end{aligned}\right\} \text{mean centre is } (20.500, 19.166)$$

$$\text{Standard distance: } d_s = \left[\sum l_{ic}^2 / n\right]^{0.5} = \left[\frac{1204.66}{12}\right]^{0.5} = 10.019$$

Calculation of d_s from the separate axes:

$$\text{standard distance on } x = d_{sx} = \left[\sum (x - \bar{x})^2 / n\right]^{0.5}$$
$$= [713/12]^{0.5} = 7.708$$

$$\text{standard distance on } y = d_{sy} = \left[\sum (y - \bar{y})^2 / n\right]^{0.5}$$
$$= [491.66/12]^{0.5} = 6.401$$

$$\text{pooled measure} \quad d_s = [d_{sx}^2 + d_{sy}^2]^{0.5}$$
$$= [7.708^2 + 6.401^2]^{0.5} = 10.019$$

method, the two axes have slightly different dispersions at 7.708 and 6.401 units. These can be used to define and plot a standard distance *ellipse* for the pattern. Just as we could generalize the mean centre to deal with point-located interval or ratio data, it is of course also possible to weight the standard distance formula in a similar way.

Distances to nearest neighbour. A third distance-based measure used to characterize point patterns is the mean of the distances between each point and its nearest, or second, third and higher-order neighbours. Most work to date has only examined the nearest-neighbour distances, and it is these we shall consider here. The procedure is very simple. Taking each point in turn, the point nearest to it is found, then the distance either calculated, or measured. Measurement is usually quick and the correct nearest point more often than not is obvious, but for accurate work, it is safest to calculate the distances from the co-ordinates. Table 3.3 sets out the computations for the point pattern of Figure 3.6. It can be seen that nearest-neighbour distances range from 4 to 11.18 units, with a mean at 6.625. Nearest-neighbour distances have been

Table 3.3 Nearest-neighbour distances for the point pattern shown in Figure 3.6(c)

Point	*Co-ordinates* (x)	(y)	*Nearest neighbour*	*Co-ordinates of nearest neighbour* (x)	(y)	l_{12} *distance*
A	5	15	C	15	10	11.18
B	10	30	G	20	25	11.18
C	15	10	E	20	15	7.07
D	16	20	F	20	20	4.00
E	20	15	H	25	15	5.00
F	20	20	D	16	20	4.00
G	20	25	F	20	20	5.00
H	25	15	I	25	20	5.00
I	25	20	K	25	15	5.00
J	30	10	H	25	15	7.07
K	30	20	I	25	20	5.00
L	30	30	K	30	20	10.00
						79.50

Mean distance to first neighbour, $l = 79.50/12 = 6.625$

much used in geographical work, but are subject to a number of obvious limitations. First, they do not take into account the direction of each point from its neighbour; if only first-order neighbours are considered, we lose a lot of information. Second, severe edge-effects may be introduced in studies where the mapped point pattern is merely a 'window' laid over a real-world pattern that extends well beyond the mapped area. Points close to the map border are forced to find neighbours within the mapped area, whereas in the real world the true nearest neighbours could lie outside the map. Finally, as described the method deals only with nominal point data. Attempts have been made to attend to all these problems: Dacey (1962, 1963) has described techniques that examine distances to higher-order neighbours that can be used with point-located interval and ratio data, while Getis (1964) outlines a method in which the nearest neighbours in direction sectors around each point are measured.

Distance counts. The final distance-based measure combines both distance and density into a useful analytic device. The method seems to be unnamed in the literature; we shall refer to it as distance counting. The method is simple: around each point symbol we draw a circle of specified radius and count the number of dots, or the total of all the values in a proportional symbol map, enclosed within it. Figure 3.7 shows the results of this operation on the dots in Figure 3.1, using circle radii of 15 and 30 distance units. The resulting maps give a very interesting new perspective on the original dot distribution. The choice of radius obviously affects the final result and, at first sight, may seem arbitrary; but, in a practical problem, it could be chosen to reflect some meaningful threshold, as for example the average distance that people might be prepared to walk to church.

Two extremely interesting features of these maps are worth examining. First, each point value can be given a physical interpretation as a *density!* It represents a number per unit of area. In Figure 3.7(a) this area is $\pi l_r^2 = 3.142 \times 15^2 = 707$ distance units^2, in (b) in the figure it is $3.142 \times 30^2 = 2827$ distance units^2. In effect, we have created a density map out of a dot distribution. The second feature is even more interesting. Although we started with a discontinuous, point-valued distribution and only centred our search circles on existing data points, there is nothing to stop us from performing counts at any point on the mapped area. There

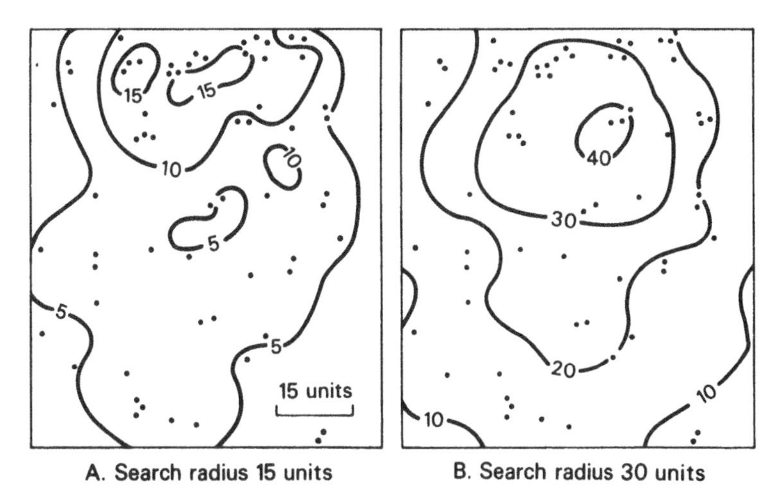

Figure 3.7 a and b Distance counts: churches in south Leicestershire.

is, therefore, a *continuous* density *surface* associated with the point distribution that can be represented by isolines of equal surface height.

Measures based on distance and direction

The methods of characterizing pattern outlined so far have numerous limitations. Quadrat counts give no information on the relative location of individuals, or on pattern at a scale below that of the grid, and are very dependent on quadrat size and shape. Distance measures neglect directional information and give severe edge-effects. To overcome some of these problems, a method that uses both distance and direction has been proposed by Vincent *et al.* (1976). Every point in the pattern has a neighbourhood assigned to it, as follows. First, find all those points that are nearer to the point in question than any other; notice that this definition of 'neighbour' will take direction as well as distance into account, and that points will have a varying number of such neighbours. Second, each of the neighbour points is joined to the central point and, third, the bisectors of all these lines are joined to define a polygonal neighbourhood around it. Such a unique partitioning of an area into neighbourhoods has been used in rainfall analysis, when it is called a *Thiessen net* after its originator. The method is illustrated in Figure 3.8. Point A has only four neighbours closer to it than any

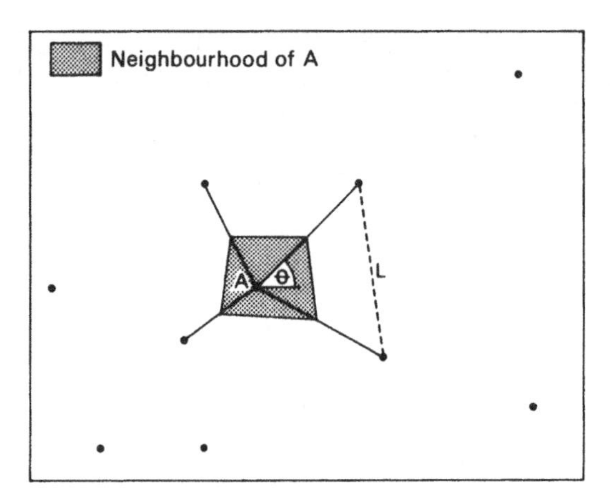

Figure 3.8 The graph-theoretic approach to pattern description.

other, giving the indicated four-sided neighbourhood. Finally, we can make use of these polygons, and their associated network of lines, to develop a number of pattern descriptors, including:

(a) n_q, the number of neighbours of a typical point or 'contact number'. In our example $n_q = 4$, but in a complete analysis we would derive a frequency distribution of these values, from $n_q = 3$ upwards;
(b) l_q, the 'link lengths' between typical pairs of neighbours;
(c) θ_q, the angles at the vertex of a typical triangle within the neighbourhood (see Figure 3.8).

Many other measures could be developed.

PROCESSES THAT GENERATE POINT PATTERNS: THE INDEPENDENT RANDOM PROCESS

Traditionally, a point symbol map was regarded as the unique result of a generating process. However, here such a map is treated as a single *realization* of a stochastic generating process, In this view, no matter what the individual unique circumstances, each point symbol is regarded as being located by a *random selection* from an underlying fixed probability distribution. If this is the case, the same random process could generate many such realizations, so that in a sense any particular map pattern is a 'sample' of size one from an underlying distribution of possible realizations. This *stochastic,* or chance, view of the map is critical to much of the content of this book, and to spatial analysis in general. The principles involved can be demonstrated readily by repeated generation of dot maps, using as the underlying distribution a table of random numbers from 0 to 99 as co-ordinates of the points; it will rapidly become evident that although the process and probability distributions are the same, very different maps can be produced (compare your results with those from exercise 3 in the Worksheet in Chapter 1 (p. 14)). Before moving on to examine some plausible probability distributions, it is important to be clear on two issues. First, note that we have used the word 'random' to describe the *method* by which the symbols are located, *not the patterns which might result*. It is the process which contains the random element, not the distribution. Second, in no sense is it asserted that spatial patterns are ultimately chance affairs. In the real world, each point symbol on the map, be it a factory or an oak tree, has a good behavioural or environmental reason for its location. All we are saying is that in aggregate these many individual histories and circumstances may best be described by regarding location processes as stochastic.

Of the great many processes that can generate point symbol maps, the simplest is one in which no spatial constraints operate, called the *independent random process*. Such a process might be appropriate were the point locations not influenced by distance, the varying quality of the environment, and so on. It is given by postulating two conditions, as follows.

(1) The condition of *equal probability:* this states that any point has equal probability of being in any position or that each small subarea of the map has equal chance of receiving a point.

(2) The condition of *independence:* this states that the positioning of any point is independent of that of any other point.

It is evident that even with these conditions operating as constraints, our spatial process could generate many diverse realizations, giving a distribution of possible maps. In order to compare what we observe on our real-world maps with the patterns that this process can generate it is necessary to be able to predict these outcomes, and to do this requires a little mathematics. Although this might seem difficult, the mathematical argument is worth following for two reasons. First, it leads eventually to a derivation of the Poisson distribution formula much used in spatial analysis and, second, it makes clear the nature of the assumptions made in applications of significance tests based on this distribution.

The mapped area is subdivided into k small subregions, called

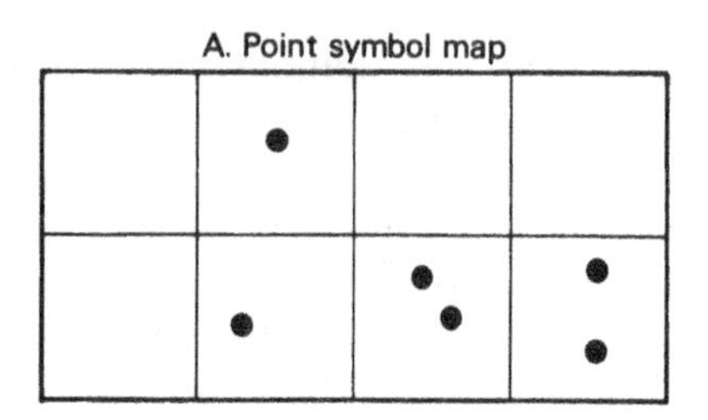

B. Frequency distribution of quadrat counts

No. of points:	0	1	2	3	4	5	6
No. of quadrats:	4	2	2	0	0	0	0

Figure 3.9 a and b A point symbol map and frequency distribution of quadrat counts.

quadrats, each of area *a*. *N* point symbols are then located by the independent random process into this area. The resulting 'map', with $k = 8$ subregions and $N = 6$ points might look something like Figure 3.9(a). We can summarize this distribution by a count—called a quadrat census—of the frequencies as outlined on pp. 38–40. But this is only *one* realization of the process. Can we say anything about *all* the realizations that the process can give? To do this, needs two further building-blocks, the *probability* that any quadrat will contain $m = 0, 1, 2, \ldots, N$ points, and the *number of ways this will result* from *N* placements.

If there are *k* subregions, the probability *p* of any point being placed in any specified region follows from condition (1) as

$$p = 1/k, \text{ or } 1/8 = 0.125 \text{ in our example.}$$

The converse, the probability of not receiving a point *q*, will be $q = (1 - p)$, or 0.875 in our example. From condition (2), we can repeat our point placements $N = 6$ times for each of our points without changing *p* and *q*, so that the probability of an area receiving just *one* point will be

$$p \times q \times q \times q \times q \times q = p^1q^5$$

This result can be achieved in any of six ways. The first dot to be placed could fall into the first quadrat, giving a realization coded

1 0 0 0 0 0

Equally, the placement might be on the second, third, etc. times

0 1 0 0 0 0
0 0 1 0 0 0
0 0 0 1 0 0
0 0 0 0 1 0
0 0 0 0 0 1

There are six ways that this could occur, so our total probability of quadrats receiving one point is $6 \times p^1q^5$. Since $p = 0.125$ and $q = 0.875$, this gives $6 \times 0.125 \times 0.875^5 = 0.3846816$. Similarly, the probability of receiving *two* points will be $p \times p \times q \times q \times q \times q = p^2q^4$, which can be achieved in fifteen different ways:

1 1 0 0 0 0	0 1 0 0 1 0
1 0 1 0 0 0	0 1 0 0 0 1
1 0 0 1 0 0	0 0 1 1 0 0
1 0 0 0 1 0	0 0 1 0 1 0
1 0 0 0 0 1	0 0 1 0 0 1
0 1 1 0 0 0	0 0 0 1 1 0
0 1 0 1 0 0	0 0 0 1 0 1
	0 0 0 0 1 1

Hence, our total probability is $15 \times 0.125^2 \times 0.875^4 = 0.1373862$. The probabilities for 0, 3, 4, 5 and 6 and for other p, q and N can be found in a similar way but we can shortcut this rather laborious process by using a little mathematics.

First, if we are interested in the probability of getting m points in a cell for $m = 0, 1, \ldots, N$, the complementary event, no points in a cell, must occur $N - m$ times, each with probability q. Hence our general formula will be $p^m q^{N-m}$ with p, q, N and m defined as before. Second, we need to know how many times each result could occur, given by the combinatorial expression:

$$\binom{N}{m} = \frac{N!}{m!(N-m)!}$$

This introduces some unfamiliar symbols. The bracketed term on the left is simply a shorthand way of saying the number of ways of getting m from N, which is what we require. The ! symbol denotes the *factorial* of the number, defined as the product

$$N! = N(N-1)(N-2), \ldots, 1$$

'Factorial 5', for example, is $5! = 5 \times 4 \times 3 \times 2 \times 1 = 120$. Notice that, by definition, $0! = 1$. Putting these two general formulae together, gives us the appropriate probability function – called the binomial – for determining the probabilities that will result from an independent random process as:

$$\text{probability } (m; N, p) = \binom{N}{m} p^m q^{N-m}$$

$$= \frac{N! p^m q^{N-m}}{m!(N - m)!}$$

The left-hand side of this equation is shorthand for the probability of there being m points in a quadrat, given a total of N and a probability p of any one point being allocated to a quadrat. The right-hand side is simply the probability of getting m points in a quadrat multiplied by the number of ways by which this could occur. It is very important to realize that these are very much long-run probabilities and that our process could generate cell frequencies that appear to have been controlled by a totally different probability distribution. Table 3.4(a) evaluates the full set of probabilities for $m = 1, \ldots, 6$ in our example, using $k = 8$ subregions and $N = 6$ points. Notice that the predicted probabilities sum to one, and that the expected long-run average number of points can be found by multiplying these probabilities by the number of points, in this case 6. It will be observed that for any specified values of k, N and m, these probabilities are uniquely determined, each depending only on the number of points in the quadrat, N the total number and p the probability of a cell receiving a specified point.

In practice, use is not often made of these binomial probabilities. They are laborious to calculate, and for most applications a second probability distribution, the Poisson, gives a more readily calculated approximation to the probabilities obtained in an independent random process. Usually in any practical application each quadrat has a small area, implying that k is large and that p is therefore small. The probability of a quadrat receiving a point is small but, by the same token, N the number of points, is usually large so that the product Np, which is the expectation of finding one point in one area, is roughly constant. If this product is called lambda, (λ), we have the Poisson distribution, given as:

$$\text{probability } (m; \lambda) = \frac{\lambda^m}{m!} \exp(-\lambda)$$

$$\lambda = Np = \frac{N}{k}$$

Table 3.4 Predicted binomial and Poisson probabilities for an independent random process, with $N = 6$, $k = 8$ and $p = 0.125$

(a)*Binomial*

m	$\binom{N}{m}$	p^m	q^{N-m}	$\binom{N}{m}p^m q^{N-m}$	*Expected no.*
0	1	1.0	0.4487952	0.4487952	2.69
1	6	0.125	0.5129088	0.3846816	2.31
2	15	0.015625	0.5861815	0.1373862	0.82
3	20	0.0019531	0.6699218	0.0261684	0.16
4	15	0.0002441	0.765625	0.0028033	0.02
5	6	0.0000305	0.875	0.0001601	0.00
6	1	0.0000038	1.0	0.0000038	0.00
				0.9999996	6.00

(b)*Poisson*

$$\lambda = 6/8 = 0.75$$

$$\exp(-\lambda) = 2.71828^{-0.75} = 1/2.71828^{0.75} = 0.4723$$

m	λ^m	$m!$	$\lambda^m \exp(-\lambda)/m!$	*Expected no.*
0	1	1	0.472300	2.83
1	0.75	1	0.354225	2.13
2	0.75^2	2	0.132834	0.80
3	0.75^3	6	0.033209	0.20
4	0.75^4	24	0.006227	0.04
5	0.75^5	120	0.000934	0.01
6	0.75^6	720	0.000117	0.00
			1.000000	6.00

The proof of this formula is given in Rogers (1974, p. 3) and, in a readily followed but rather different form, in Lewis (1977, pp. 70–4). It requires a little explanation. We have seen the factorial $m!$ before, but $\exp(-\lambda)$ needs further comment. The symbol exp refers to the number e = 2.71828 . . . , which often turns up in mathematics as the sum of the infinite series:

$$e = \frac{1^0}{0!} + \frac{1^1}{1!} + \frac{1^2}{2!} + \frac{1^3}{3!} + \ldots$$
$$= 1 + 1 + 0.5 + 0.1667 + \ldots = 2.71828 \ldots$$

The term exp(-λ) is, thus, found from $1/2.71828^{\lambda}$, and the predicted probabilities according to this Poisson distribution are found from the expression given above. Table 3.4(b) shows the complete Poisson probabilities for our example with $\lambda = 6/8 = 0.75$. It can be seen that, even with these low values for N and k, the probabilities are not all that different from those given by the binomial.

The importance of these results cannot be overstated. In effect, we have specified a process – the independent random process – and have used some mathematics to predict the frequency distribution that, in the long run, quadrat counting of its realizations should yield. These probabilities can, therefore, be used as a standard by which any observed real-world distribution can be judged. The analysis has been conducted for probabilities that can be used in a quadrat analysis of pattern, but similar arguments can be used to predict expected values of other pattern descriptors. In their paper of 1954, Clark and Evans show, for example, that if the distribution of nearest-neighbour distances is normal, the independent random process will yield an expected, or long-run average, mean distance to nearest neighbour of

$$\bar{l}_e = 0.5/(d)^{0.5}$$
$$d = N/a$$

the overall density (N = number of points, a = area enclosing them). Moreover, this average has a standard error of

$$S_{l_e} = 0.26136/(Nd)^{0.5}$$

For the mathematically inclined a readily accessible derivation is given in Rogers (1974, p. 8). Less is known about the expected values and distributions of some of the other measures proposed on pp. 47–8 (see Boots, 1977). The expected distribution of the number of neighbours defined using Vincent's approach has been obtained not by mathematical means, but by a simulation in which a large number – 1000 – of point patterns was generated using the independent random process, then simply counting the neighbours. While not mathematically exact, the proportions of the total at each value of n_q listed in Table 3.5 can be used as a reasonable approximation. Geometricians and geographers interested in central place theory will be relieved to note that the most probable polygon is a hexagon!

Table 3.5 Frequencies of the contact numbers of neighbours in Thiessen nets, generated by an independent random process (after Vincent *et al.* 1976)

n_q	3	4	5	6	7	8	9	10	11	12
% total	1.06	11.53	26.47	29.59	19.22	8.48	2.80	0.66	0.12	0.03

TESTING OBSERVED PATTERNS AGAINST THOSE GENERATED BY THE INDEPENDENT RANDOM PROCESS

The two previous sections in this chapter contained the building-blocks that enable the spatial analyst to use the power of statistical hypothesis-testing in his work. In this section it is assumed that the reader is familiar with the fundamentals of statistical-testing as outlined in any of the standard texts (see for example Hammond and McCullagh, 1978, chapter 6). A number of statistical tests may be used in point pattern analysis, and the choice will depend upon the measures of pattern that have been calculated. Here, we discuss four tests that are appropriate in analysing point symbols, but almost all the standard tests have complications when applied to spatial data. The tests are Student's t-test, using the observed and expected variance/mean ratios, the chi-square and Kolmogorov-Smirnov tests on quadrat counts, and the nearest-neighbour test based on distances.

The variance/mean ratio test

It can be shown readily that the mean and the variance of a histogram of quadrat counts derived from a pattern generated by the Poisson process are equal, so that their ratio should have the value 1. Accordingly, we can set up the hypothesis that an observed variance/mean ratio (VMR), such as the 1.5903 derived in Table 3.1 for the pattern of churches, is drawn from a population whose mean is unity. A measure of departure of observed from expected is simply their difference, and it can also be shown that the variance/mean ratio has a standard error given by

$$S_{\text{VMR}} = [2/(k\text{-}1)]^{0.5}$$

where k = number of cells, which enables a Student's t-value to be calculated as

$$t = \frac{\text{observed VMR} - 1}{\text{standard error of difference}} = \frac{\text{observed VMR} - 1}{[2/(k - 1)]^{0.5}}$$

and tested in the usual way with k - 1 degrees of freedom. For the church data we have

$$t = \frac{1.5903 - 1}{[2/(56 - 1)]^{0.5}} = \frac{0.5903}{0.1907} = 3.0956 \text{ with } 55\ df$$

The probability of such a t-value or one larger arising by chance is less than 1 per cent. *Either* our distribution of churches is an extremely rare realization of the independent random process, *or* the null hypothesis (H_0) is unacceptable. Conventionally, we must reject H_0 and conclude that the generating process is not independent random. As outlined on p. 39, there appears to be a distinct clustering tendency.

The chi-square and Kolmogorov-Smirnov tests

The variance/mean ratio method is a parametric test that involves calculating two parameters (the mean and variance) of the frequency distribution of quadrat counts. The chi-square and Kolmogorov-Smirnov tests are non-parametric alternatives that operate directly on the observed frequencies. In Figure 3.5(b) and Table 3.1 we saw the counts of the observed frequencies of quadrats containing $m = 0, 1, \ldots, 5$ points. If we set up the hypothesis that these were generated by an independent random process, then the expected frequencies given by the Poisson distribution may be used in either a chi-square or a Kolmogorov-Smirnov D-test. The relevant computations for the chi-square analysis are laid out in Table 3.6, giving a chi-square value of 5.531 which with three degrees of freedom is significant at the 80 per cent level, but not at 90 per cent. Several problems are apparent when using this test. First, the well-known sensitivity of the statistic to low expected frequencies has been avoided by combining all frequencies of three or more points to give expected frequencies all greater than 5, the usually accepted critical value for a valid test (but see Rogers, 1974, p. 68). Second, the analysis seems to confirm Mead's (1974) scepticism as it fails to detect a genuine difference between observed and expected as detected by Student's t-test.

An alternative test that avoids the problem of low frequencies and which does not lose as much information is the Kolmogorov-Smirnov D-test, which uses the maximum difference between observed and predicted cumulative frequency distributions. The relevant calculations are laid out in Table 3.7, which shows that we

Table 3.6 Chi-square analysis for the quadrat-count data for churches in south Leicestershire (Figures 3.1 and 3.5)

Expected frequencies

number of cells $= k = 56$

number of points $= n = 60$

$\lambda = n/k = 60/56 = 1.07\dot{1}4$

$\exp(-\lambda) = 0.3426$

m	$\lambda^m \exp(-\lambda)$	*m!*	*Probability*	*Expected no.*
0	0.3426	1	0.3426	19.186
1	0.3671	1	0.3671	20.557
2	0.3933	2	0.1966	11.010
⩾3	residual calculated		0.0937	5.247
			1.0000	56.000

Calculation of test statistic

m	*Observed*	*Expected*	$(O\text{-}E)^2/E$
0	25	19.186	1.762
1	15	20.557	1.502
2	8	11.010	0.823
⩾3	8	5.247	1.444
	56	56.000	5.531

Degrees of freedom = number of classes − 1 = 4 − 1 = 3

Table 3.7 Kolmogorov-Smirnov analysis for the quadrat-count data for churches in south Leicestershire (Figures 3.1 and 3.5)

m	*Observed cell Count*	*Observed cumulative proportion*	*Expected cumulative proportion*	*Difference*
0	25	0.446	0.343	0.103
1	15	0.714	0.710	0.004
2	8	0.857	0.906	−0.049
3	5	0.946	0.977	−0.031
4	1	0.964	0.995	−0.031
⩾5	2	1.000	1.000	0.000

cannot reject the null hypothesis of an independent random generating process at the 95 per cent level. This test is an attractive one, in that it avoids a lot of the computational labour associated with Student's *t*-test and the chi-square, but unfortunately published tables of critical values of the test statistic D are based on the assumption that the expected distribution is specified before the analysis, that is with controlling parameters which are not derived from the same data as were used in obtaining the observed frequencies. In our analysis we estimated λ from these data, so that strictly speaking the test is invalid. Although the details are not covered here, it is also possible to use these tests on the frequency distributions of the indices proposed by Vincent *et al.* (1976); the computations for a simple example are laid out in their paper.

The nearest-neighbour index

The final test uses the mean distance to nearest neighbour $\bar{l}$, but like the previous tests it is still based on what would be expected to occur in the long run if the generating process were controlled by a Poisson distribution. On p. 54, we noted that the expected distance

$$\bar{l}_e = 0.5/(d)^{0.5}$$

so that we can use the ratio of the actual observed mean distance to this random expectation as a means of characterizing the pattern relative to this 'random' standard. This is the nearest-neighbour index, defined as:

$$R = \bar{l}/\bar{l}_e$$

giving a value for 1 for a Poisson process, values of less than 1 a tendency to cluster, and more than 1 a tendency towards uniform spacing. We can calculate this R statistic for the distribution of Leicestershire churches (Figure 3.1) as follows. Using the (x, y) co-ordinates of the points, the observed mean distance to nearest neighbour is found to be 3.177, and the expected

$$\bar{l}_e = 0.5/(60/5600)^{0.5} = 4.830$$

(Notice that the area required for the calculation of the density d must be in 'squared' units of the unit used to measure distance.) The nearest-neighbour statistic is therefore

$$R = 3.177/4.830 = 0.658$$

indicating the same tendency to cluster detected by the variance/mean ratio test.

Nearest-neighbour statistics are in themselves useful as descriptive devices and have been much used in the literature. What is less frequently seen is a significance test on the distances but this can be accomplished without much extra effort. The standard deviation of l_e is given by

$$s_{l_e} = 0.26136/(Nd)^{0.5}$$

This enables us to form a standard normal deviate, or z-score, as:

$$z = (i - l_e)/s_{l_e}$$

and to find the probability of z from tables of the normal distribution with the null hypothesis that $z = 0$, the mean distance is identical to that expected. Notice that in performing the test, we make a further assumption (which might not be justified in a practical application), that the differences are normally distributed. Performing the arithmetic for the church data yields

$$s_{l_e} = 0.26136/(60 \times 60/5600)^{0.5} = 0.326$$

$$z = (3.177 - 4.830)/0.326 = -5.071$$

The probability of such a high z-score occurring by chance with a true null hypothesis is much less than 0.1 per cent, so we infer that the pattern is unlikely to have been generated by a Poisson process; the distribution is not described by an independent random model. As geographers, we are now in a position to speculate as to why this is the case, and almost certainly we should next examine the distribution of population in this area.

In this section, we have shown how some of the pattern descriptions developed on pp. 36–48 can be used to test for significance using as null hypothesis the pattern measures that would be generated by an independent random process. It is of interest to note that, depending on the method used, we have obtained slightly different results. The variance/mean ratio and nearest-neighbour tests both lead us to reject the null hypothesis that the distribution of churches in south Leicestershire was generated randomly, whereas the chi-square and Kolmogorov-Smirnov tests would not allow the null hypothesis to be rejected at a conventionally high level of, say, 95 per cent. Logically, there is no conflict between these results, but they should warn against the casual use of a single test conducted mechanically. The statistical analysis is not, and never can be, a valid substitute for careful thought. The tests described are the widest used, but in the statistical literature many others have been described; the interested reader is referred to work by Greig-Smith (1964), and Mead (1974).

PROCESSES THAT GENERATE POINT PATTERNS: DEPENDENCE IN SPACE

The independent random process and some tests have been outlined that enable observed point patterns to be tested against the realizations of the process. This independent random process is mathematically elegant, relatively simple and forms a useful starting-point for spatial analysis, but in geography its use is often exceedingly naïve and unrealistic. Most geographical applications of this model have been made in the hope of being forced to reject the null hypothesis of independence and randomness in favour of some alternative hypothesis that postulates a spatially dependent process. Were real-world point patterns generated by unconstrained randomness, then 'geography' as we understand it would have little meaning or interest, and examination of most maps will rapidly suggest that some other process is operating. In the real world events at one place and time are only seldom independent of events at another, so that as a general rule we expect point patterns to display spatial dependence. Two major possibilities can be outlined. Consider, for example, the settlement of the Canadian prairies in the latter half of the nineteenth century. As settlers spread, market towns grew up in competition with one another. For various reasons, notably the competitive advantage given by being on a railway, some towns prospered while others declined, with a strong tendency for successful towns to be located as far as possible away from other towns. The result is spatial separation and a tendency to give uniform spacing of the sort predicted by central place theory. Other real-world processes involve the idea of aggregation or clustering, in which the act of locating one point increases the probability of others being located close to it. Examples might include the spread of a contagious disease like the 1967 outbreak of foot and mouth disease in England, the diffusion of an innovation through an agricultural community, or, more simply, where variations in the quality of the environment lead to clustering of mines, houses, factories, or whatever in the most favoured areas.

These departures from an independent random model could be *detected,* using the tests outlined on pp. 55–60, but can we model the precise nature of processes that incorporate spatial dependence? Clearly, to give uniformity or clustering requires that the probability of any small subarea receiving a point is not the same for all cells or all of the time. A number of models that incorporate spatial dependence have been suggested and evaluated, but the precise mathematical details are well beyond the scope of an elementary text such as this.

Suppose that we postulate a *contagion* process, in which the probability of a placement increases linearly with the number of points already in the cell, leading fairly directly to a clustered point pattern. Rogers (1974, p. 16) shows that this yields a frequency distribution of quadrat counts that can be described by a Negative Binomial probability distribution. A second process we might envisage involves the random placement of a series of 'parents', from each of which the offspring spread randomly. Because the resulting distribution was first derived by Neyman to describe the distribution of larvae around a random distribution of egg clusters, it is referred to as the Neyman Type A. Processes that tend towards uniformity of distribution have been less well studied, the notable exception being a series of models proposed by the geographer and mathematician, M. F. Dacey (see Dacey, 1960, 1963, 1964, 1966; Dacey and Tung, 1962).

Whatever the process, each leads to a probability distribution of quadrat counts, which can be tested against the observed point patterns precisely in the same way as outlined on pp. 55–60. We compare observation with model prediction and decide how likely the observations are. The methodological difficulties involved are many and varied and are summarized in a series of papers by Harvey (1966, 1967). First, the derivation of appropriate probability models to match even the simplest postulated spatial process is not simple, requiring a thorough knowledge of probability theory. Second, in most geographical applications it is usual to estimate the parameters that control the model from the observed data. The predicted frequencies from such an *a posteriori* model are then tested against the observations, hence it is likely that a reasonable 'fit' will be obtained. Even in the simplest Poisson model, remember that the parameter λ was *estimated* from the data as N/k. If our data had truly arisen from a quadrat sample, this value itself is an estimate of some unknown underlying population value. In short, there is a circularity of argument which ensures that a good 'fit' of theory to observation does not necessarily imply that we have identified the correct process model. Despite these difficulties, it is apparent that this type of analysis, in which a process is postulated, modelled as a probability model and then compared with observation, will gain increasing application in spatial analysis.

RECOMMENDED READING

·There is a large and rapidly expanding literature on the mapping and analysis of point symbol symbols.

Cartographic considerations

Bachi, R. (1968) *Graphical Rational Patterns,* Jerusalem.

Cox, C. W. (1976) 'Anchor effects and the estimation of graduated circles and squares', *American Cartographer* 3, 65–74.

Dahlberg, R. E. (1967) 'Towards the improvement of the dot map', *International Yearbook of Cartography* 7, 157–67.

de Geer, S. (1922) 'A map of the distribution of population in Sweden: method of preparation and general results', *Geographical Review* 12, 72–83.

Dixon, O. (1979) 'A map of population distribution in the Portsmouth area', *Cartographic Journal* 16, 9–13.

Dobson, M. W. (1974) 'Refining legend values for proportional circle maps', *Canadian Cartographer* 11, 45–53.

Flannery, J. J. (1971) 'The relative effectiveness of some common graduated point symbols in the presentation of quantitative data', *Canadian Cartographer* 8, 96–109.

Mackay, J. R. (1949) 'Dotting the dot map', *Surveying and Mapping* 9, 3–10.

Mackay, J. R. (1953) 'Percentage dot maps', *Economic Geography* 29, 263–6.

Monmonier, M. S. (1977) 'Regression-based scaling to facilitate the cross-correlation of graduated circle maps', *Cartographic Journal* 14, 89–98.

Olson, J. M. (1975) 'Experience and the improvement of cartographic communication', *Cartographic Journal* 12, 94–108. See also Williams (1977).

Robinson, A. H. and Sale, R. D. (1969) *Elements of Cartography,* 3rd edn, New York, Wiley, 115–31.

Stewart, J. Q. (1947) 'Empirical mathematical rules concerning the distribution and equilibrium of population', *Geographical Review* 37, 461–85.

Williams, R. K. (1977) 'Correspondence: graduated circles; *Cartographic Journal* 14, 80.

Williams, R. L. (1956) *Statistical Symbols for Maps: Their Design and Relative Values,* Yale University, Map Laboratory.

Statistical problems: texts and summaries

There are a number of excellent texts that explore point pattern analysis in great detail. Recommended are:

Getis, A. and Boots, B. (1978) *Models of Spatial Processes,* Cambridge, Cambridge University Press.

Greig-Smith, P. (1964) *Quantitative Plant Ecology*, London, Butterworths.
Haggett, P., Cliff, A. D. and Frey, A. (1977) *Locational Methods*, London, Arnold, especially 414–47.
Hammond, R. and McCullagh, P. S. (1978) *Quantitative Techniques in Geography: An Introduction*, 2nd edn, Oxford, Clarendon.
Lewis, P. (1977) *Maps and Statistics*, London, Methuen.
Rogers, A. (1974) *Statistical Analysis of Spatial Dispersion*, London, Pion.
Thomas, R. W. (1977) *An Introduction to Quadrat Analysis*, Norwich, Geo Abstracts, CATMOG, 12.

Research papers on specific methods

Boots, B. (1977) 'Comments on "Urban settlement patterns and the properties of the Simplicial graph" by P. Vincent, J. Haworth, J. Griffiths and R. Collins', *Professional Geographer* 29, 411–12.
Clark, P. J. and Evans, F. C. (1954) 'Distance to nearest neighbour as a measure of spatial relationships in populations', *Ecology* 35, 445–53.
Dacey, M. F. (1960) 'The spacing of river towns', *Annals, Assoc. Amer. Geogr.* 50, 59–61.
Dacey, M. F. (1962) 'Analysis of central place and point patterns by a nearest neighbour method', in Norborg, K. (ed.) *Proceedings, IGU Symposium on Urban Geography*, Gleerup, Lund Studies in Geography, B, 24.
Dacey, M. F. (1963) 'Order neighbour statistics for a class of random patterns in multidimensional space', *Annals, Assoc. Amer. Geogr.* 53, 505–15.
Dacey, M. F. (1964) 'Modified Poisson probability law for a point pattern more regular than random', *Annals, Assoc. Amer. Geogr.* 54, 559–65.
Dacey, M. F. (1966) 'A county seat model for the areal pattern of an urban system', *Geographical Review* 56, 527–42.
Dacey, M. F. (1967) 'Two dimensional random point patterns: a review and interpretation', *Papers, Regional Science Association* 13, 41–55.
Dacey, M. F. and Tung, T. (1962) 'The identification of randomness in point patterns', *Journal of Regional Science* 4, 83–96.
Getis, A. (1964) 'Temporal land use pattern analysis with the use of nearest neighbour and quadrat methods', *Annals, Assoc. Amer. Geogr.* 54, 391–9.
Greig-Smith, P. (1952) 'The use of random and contiguous quadrats in the study of the structure of plant communities', *Annals of Botany* n.s. 16, 293–312.
Harvey, D. W. (1966) 'Geographical processes and the analysis of point patterns', *Transactions, Inst. Brit. Geogr.* 40, 81–95.
Harvey, D. W. (1967) 'Some methodological problems in the use of Neyman Type A and negative binomial distributions for the analysis of spatial point patterns', *Transactions, Inst. Brit. Geogr.* 44, 85–95.
Mead, R. (1974) 'A test for spatial pattern at several scales using data from a grid of contiguous quadrats', *Biometrics* 30, 295–307.

Nystuen, J. D. (1968) 'Identification of some fundamental spatial concepts' in Berry, B. J. L. and Marble, D. F. (eds) *Spatial Analysis: A Reader in Statistical Geography,* Englewood Cliffs, NJ, Prentice-Hall.
Skellam, J. G. (1953) 'Studies in statistical ecology', *Biometrika* 39, 346–62.
Thompson, H. R. (1956) 'Distribution of distance to nth neighbour in a population of randomly distributed individuals', *Ecology* 37, 391–4.
Vincent, P. J. *et al.* (1976) 'The detection of randomness in plant patterns', *Journal of Biogeography* 3, 373–80.

WORKSHEET

(1) The evaluation of its usefulness in a practical application is one of the best ways of following an analytic technique. In geomorphology a number of workers have attempted to use point pattern analysis to understand the processes that give rise to drumlin fields, as for example:

Smalley, I. J. and Unwin, D. J. (1968) 'The formation and shape of drumlins and their distribution and orientation in drumlin fields', *Journal of Glaciology* 7, 377–90.
Trenhaile, A. S (1971) 'Drumlins, their distribution and morphology', *Canadian Geographer* 15, 113–26.
Trenhaile, A. S. (1975) 'The morphology of a drumlin field', *Annals, Assoc. Amer. Geogr.* 65, 297–312.
Crozier, M. J. (1976) 'On the origin of the Peterborough drumlin field: testing the dilatancy theory', *Canadian Geographer* 19, 181–95.
Rose, J. and Letzer, J. M. (1976) 'Drumlin measurements: a test of the reliability of data derived from 1:25,000 scale topographic maps', *Geological Magazine* 112, 361–71.
What can you learn from this sequence of papers?

(2) The Steinhaus paradox for irregular curves can be illustrated by the following exercise. Choose a length of coastline and use dividers set at a constant interval to measure its length by the 'step' method (i.e. simply counting the number of times this fixed length is stepped off along the line) from a number of topographic maps from 1 : 100,000

Continued on next page

down to 1:10,000. Repeat using different settings of the dividers.

Next plot a graph of the common logarithm of these measured lengths against that of the interval set on the dividers and compare your results with Mandelbrot, B. B. (1967) 'How long is the coast of Britain?', *Science* 155, 636–8, or with the same author's (1977) *Fractals: form, chance and dimension,* San Francisco, Freeman, especially chapter 2.

(3) Table 3.3 has data for the locational (x, y) co-ordinates of a point pattern, in which $n = 12$ and with a mean distance to nearest neighbour of $\bar{l} = 6.625$. Complete a nearest-neighbour analysis by computing R and its associated z-score. What is the probability of this z-score occurring by chance with a true null hypothesis of independent randomness?

(4) (This exercise needs groups of two people.) Without consulting the partner, one person should prepare or copy a graduated symbol map in which the symbol areas are made proportional to the values being represented. A key should be provided to show, say, four typical symbols and their associated values. The other person should now attempt to estimate the mapped z-values from the evidence of the circle sizes. Plot a graph of the common logarithm of the estimated values against that of the true values and compute the regression of y on x in logarithms. Compare the value of the slope obtained with the results of Williams, Flannery and Olson in the three papers cited on p. 63.

CHAPTER FOUR

·LINES ON MAPS·

INTRODUCTON

Chapter 3 outlined the cartographical and analytical problems in mapping point data having zero length dimension. This chapter deals with line data having the spatial dimension of a length L^1. Lines can be used in many ways to represent a wide variety of information: they can show paths across areas, connected trees, interconnected networks, or be used to demarcate areas and as contour lines to show surfaces. It follows that, cartographically, the distinction made in Chapter 2 between lines, areas and surfaces is somewhat arbitrary, but analytically there is a major difference between true line patterns (paths, trees and networks) and patterns of areas and surfaces which are not intrinsically linear, but which are usually represented by lines. The difference is in the phenomena studied, not in the way they are mapped, a distinction that we shall maintain in this chapter.

Imagine that you have drawn a series of lines representing roads in an area. The concept of *pattern* introduced on p. 37 for point data has equal application here; but in addition three new fundamental spatial concepts are added – *distance, direction* and *connection.* Lines have length as well as direction, and must connect at least two points, thus, these three concepts are prominent throughout this chapter.

THEORY AND PRACTICE IN LINE SYMBOL MAPPING

As we have seen, cartographers have carefully examined problems in mapping using point symbols, but far less has been written about line symbols and we know very little about them as communications devices. A start can be made by attempting a typology of all possible types of line symbol related to three fundamental characteristics,

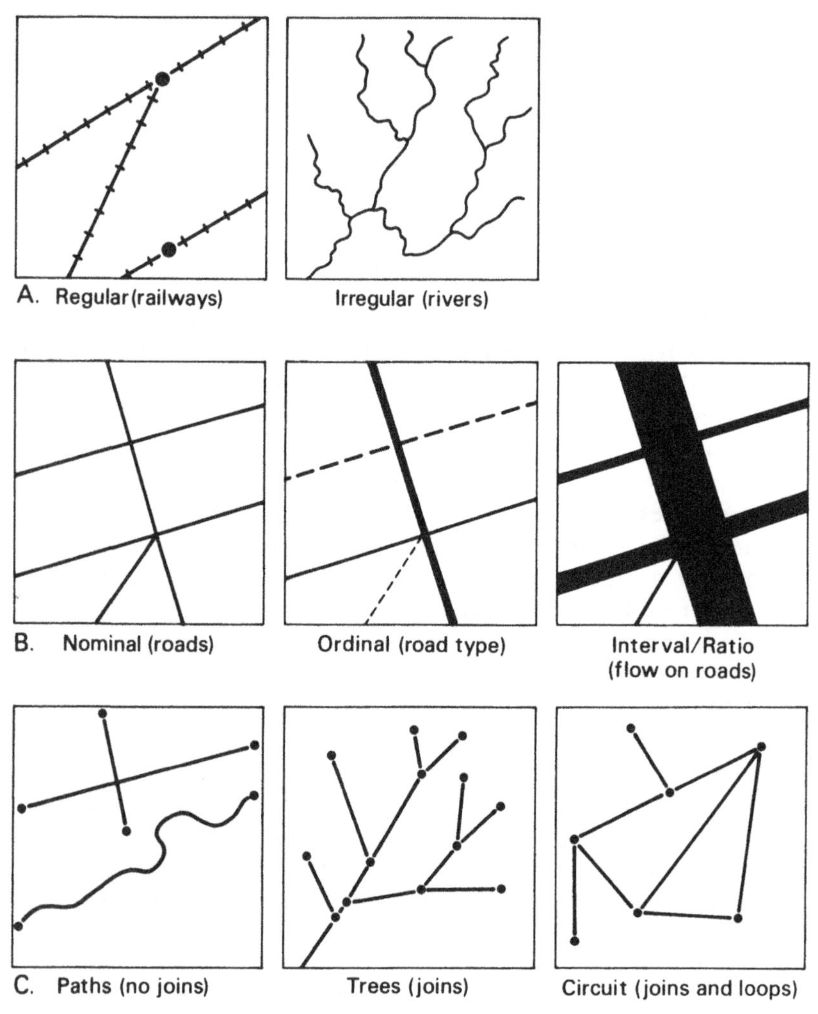

Figure 4.1 a, b and c Types of line symbol map.

whether the lines are *regular* or *irregular*, the *level of measurement* of the information they portray, and the *type of pattern* created. These distinctions are illustrated in Figure 4.1.

At first sight, the difference between a regular and an irregular line seems obvious, but on examination it turns out to be rather less clear-cut. By *regular*, we mean that the line can be described by a mathematical function, as for example the straight line described by

$$y = mx + c$$

in which x and y are spatial co-ordinates and m and c are constants. If $c = 5.0$, $m = 0.5$, when $x = 10.0$, then y will have a value of

$$y = 0.5(10.0) + 5.0 = 10.0$$

and at $x = 20$, $y = 15$. In contrast, *irregular* lines cannot be described by simple mathematical functions, an interesting consequence being the Steinhaus paradox (see Chapter 3 and exercise 2 in the Worksheet, on p. 65), that the length seems to increase, the more accurately we measure it. Some line features of the real world such as railways and canals are usually regular, others such as coastlines and watercourses are irregular, but on the map this distinction also depends on scale. For convenience and clarity, irregular lines must be generalized to a greater or lesser extent and it may even be necessary to make regular curves even more regular.

The second basis for this typology of lines on maps is that of the measurement level of the information represented. *Nominal* line data simply indicate the existence of a line feature such as a road, river or economic link, and examples from an Ordnance Survey 1 : 50,000 sheet are rivers, canals, pipelines and electricity transmission lines. *Ordinal* line data introduce a need to vary the way in which the line is drawn so as to imply an ordered measurement. For example, the same sheet has line symbols of increasing prominence to show eight different road types from minor roads to motorways. *Interval* and *ratio* line data are also represented by varying the line, usually in its width, in proportion to the value being displayed. The quantitative value associated with the line can be some characteristic of the line itself (for example, canal width, railway gauge), or it may be a measure of flow along the link (for example, traffic flow). The direction of flow can also be shown. Finally, line data can be classified by the pattern they create, as simple join-less *paths,* as connected *trees* in which joins occur but which have no closed loops, or as *circuits* in which such loops do occur.

According to this three-way typology, we have already allowed for the possibility of 18 (= 2 × 3 × 3) types of lines on maps. Table 4.1 lists some real-world examples of what might be represented in each category. It will be seen that there are categories which, in practice, are extremely hard to exemplify. The variety of line types gives the cartographer a great deal of choice in deciding which aspect of the pattern he wants to represent most clearly. Three

Table 4.1 Examples of types of line symbol map

Regular curves			
	Pattern		
Level	*Path*	*Tree*	*Circuit*
Nominal	bird migration route	commuter routes to central place	railway map
Ordinal	major/minor migration	cheap/modest/expensive modal transport split	single- or multiple-track railway
Interval and Ratio	bird flow map	commuter flows into central place	railway flow

Irregular curves			
	Pattern		
Level	*Path*	*Tree*	*Circuit*
Nominal	trace of insect paths	drainage network	?
Ordinal	large/small insect path	Strahler ordered network	?
Interval and Ratio	size of insect path	river discharges	?

issues are involved; the choice of line thickness used to show nominal data; the choice of symbolism used for ordinal data; and the method of scaling line thickness to show interval or ratio data.

First, lines must be sufficiently wide to be clear and sharp, the easiest and simplest method being to use solid, continuous lines rather than the various dotted, dashed or pecked alternatives shown in Figure 4.2. In drawing the line, care should be taken to ensure that it is of uniform width and that it will remain visible on the final map product. A practical guide to visibility is that, under normal viewing conditions, the line width must exceed that which subtends an angle of two minutes of arc at the eye, giving the list of minimum widths from different viewing distances presented in Table 4.2. Clearly, maps for wall display with a viewing distance greater than 1 m should never use linework thinner than 0.5 mm

Table 4.2 Minimum line widths at different viewing distances

Distance (*m*)	*Minimum width* (*mm*)
0.1	0.05
1.0	0.57
10.0	5.70
100.0	57.00

and, remembering that these are minimum values, there is a strong case for making them much thicker. In practice, line width is usually determined by the nib sizes available in standard sets of pens, usually from 0.25 mm to 5 mm, above which two lines are drawn and the enclosed area is shaded in.

A second consideration which applies when ordinal line data are represented, or where several types of nominal line data are shown on the same base, is the choice of line symbolism. One possibility is to use different widths to symbolize different levels or types, but this needs care. The right-hand side of Figure 4.2 shows a grid of lines of increasing width, which makes it quite obvious that, in order for the eye to be able to discriminate between them, the contrasts must be very marked. It is only in the upper right and lower left of the grid that the lines seem different. A second possibility is to alter the line symbolism or colour along some assumed graded series, but the difficulties of achieving the desired effect make this a hazardous business.

Finally, we may wish to modify the line symbol to represent inter-

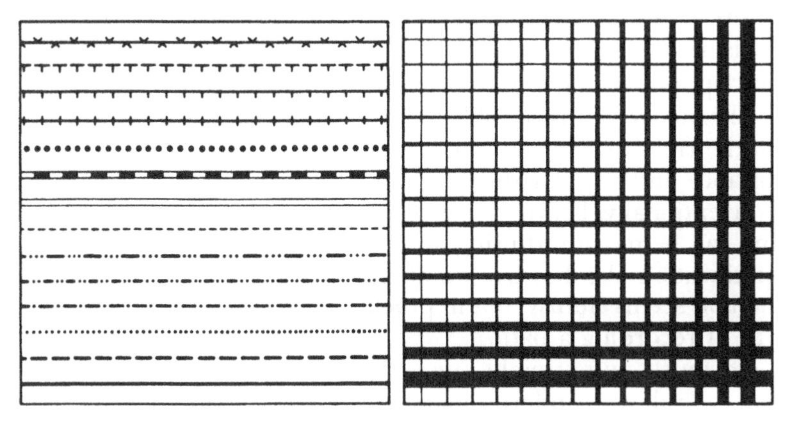

Figure 4.2 Types of line symbol and size contrast in lines.

val or ratio scaled line data. Just as the area of a point symbol can be varied in proportion to the values being mapped, so it is possible to vary line thickness in the same way. This works reasonably well where the data are constant along the entire line, and where the range of values is not extreme, but the eye has great difficulty in detecting the minor differences in width that may be necessary.

DESCRIBING LINES ON MAPS

In Chapter 3 (pp. 36–48) a number of methods for describing point patterns were described. The present section attempts to do the same for lines, beginning with the ideas of distance and direction as they apply to paths, and going on to examine measures of connection in tree and circuit networks.

Path description

The individual line or path is the basic building-block of any line pattern, and the most important measurement that we can make is of its *path length* (l). As seen on p. 41, measuring the straight-line distance between two points on a plane surface is not too difficult. As shown in Figure 4.3(a), the length of a straight-line path from an origin at (x_1, y_1) to an end-point at (x_2, y_2) is given by the Pythagoras theorem as

$$l_{1,2} = [(x_2 - x_1)^2 + (y_2 - y_1)^2]^{0.5}$$

To record such straight-line data, we need to remember only the co-ordinates of the origin and end-point, together with some measure (z) of the type of path or flow along it. The complete description has five numbers x_1, y_1, x_2, y_2, and z.

Alternatively, it is possible to record a path by the method shown on the right of Figure 4.3(a), specifying the origin followed by the distance and angular direction, or bearing θ to the second point. Conventionally, bearings are given as angles clockwise from north and can be measured using an ordinary protractor. The resultant specification also has five numbers: x_1, y_1, l, θ and z. This second method seems clumsy, but in practical navigation it is a very useful way of recording a path.

Obviously, our straight line is the shortest distance between two points and will underestimate the distance along any irregular path. To obtain the length of an irregular path, it is necessary to insert intermediate points, linking them with short, straight-line

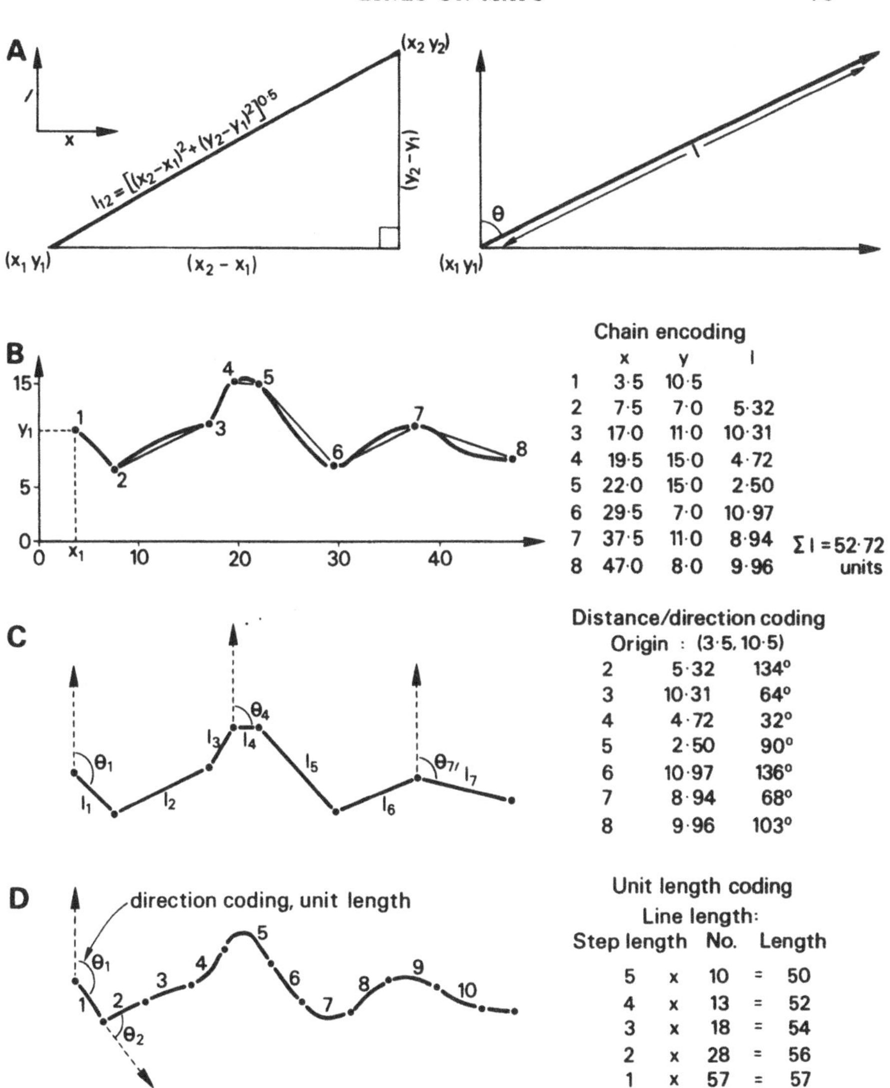

Chain encoding

	x	y	l
1	3·5	10·5	
2	7·5	7·0	5·32
3	17·0	11·0	10·31
4	19·5	15·0	4·72
5	22·0	15·0	2·50
6	29·5	7·0	10·97
7	37·5	11·0	8·94
8	47·0	8·0	9·96

Σ l = 52·72 units

Distance/direction coding

Origin : (3·5, 10·5)

2	5·32	134°
3	10·31	64°
4	4·72	32°
5	2·50	90°
6	10·97	136°
7	8·94	68°
8	9·96	103°

Unit length coding

Line length:

Step length		No.		Length
5	x	10	=	50
4	x	13	=	52
3	x	18	=	54
2	x	28	=	56
1	x	57	=	57

Figure 4.3 a, b, c and d Methods of recording paths.

segments as shown in Figure 4.3(b). In practice, there are two ways of finding the distance. The first is to mark all the important turning-points on the line, and to calculate the total as the sum of these individual lengths. As can be seen, this process of chain encoding gives a length estimate of 52.72 units. Nowadays chain encoding is not as laborious as it looks, because use can be made of a computer-controlled table digitizer. This is a special table on to which the map is fixed and lines of interest tracked using a 'follower', or cursor. As this is moved, so the device automatically produces a string of locational co-ordinates. An alternative to using co-ordinates, is to use the distance/direction method, shown in Figure 4.3(c). Notice that in both methods, and in order to maximize the fit of the short segments to the original irregular line, we have used a variable segment length. For many practical problems it is easier, quicker and cheaper to use measures based on a fixed segment length, as illustrated in Figure 4.3(d). Simply set a pair of dividers at a fixed, relatively short distance and then step along the line, counting the number of steps. The total length is this number multiplied by the individual step length. The step length should be varied according to the irregularity of the line measured and required accuracy. In Figure 4.3(d), a length of 5 units gave a length estimate of 50 units, whereas a length of 3 units gave a total of 54 units, and so on. A third advantage is that the line can be recorded very simply as the origin and step length followed by a string of angular bearings $\theta_1, \ldots, \theta_n$, a process called *unit length coding* (Baxter, 1976). Whatever length measurement procedure is used, it is apparent that care should be taken to ensure that any comparative measures are made with the same measurement method and are of an accuracy appropriate to the problem being studied.

Having measured the length of a path, or series of paths, it is possible to derive indices to describe the overall line pattern. To describe a wandering or meandering path, a simple measure is the *sinuosity ratio:*

$$S = \frac{\text{observed path length}}{\text{straight-line distance from origin to end}}$$

In the example (Figure 4.3) the straight-line distance from end to end is 43.57 units, and our most accurate measure of observed path length is 57, giving a sinuosity ratio of

$$S = 57/43.57 = 1.31$$

Notice that this number is the ratio of length L, divided by another length, and so has dimension $LL^{-1} = L^0$, that is it is a dimensionless number of the type discussed in Chapter 2 (p. 19). Despite its obvious dependence on how the lengths are measured (in our example it could be anything from 1.15 to 1.31, according to method), the ratio has been found useful in studying river channel patterns in geomorphology. In transport geography, a problem examined by Nordbeck (1964) and Timbers (1967) is the relationship between actual road distances between places and the direct, straight-line distance. Their *route factor*, or ratio of observed road distance to straight-line distance, is identical to the sinuosity ratio as defined above. Nordbeck found that in Sweden route factors averaged out at about 1.2, while Timbers found an average of 1.17 over the UK.

The defect of the sinuosity ratio is that it tells us very little about the nature of departures from straight, a single large loop giving the same value as numerous small ones. A second way of examining meandering behaviour is to examine directional changes along the path. The basic method is to isolate the change in direction from one segment of the path to the next at fixed distance intervals in the 'downstream' direction, as illustrated in Figure 4.3(d). For simplicity, consider only three possibilities, a path segment which goes straight ahead (±45° from the previous segment), a left-turning segment (more than 45° to the left) and a right-turning segment. Thus, in passing from segment 1 to segment 2, the path turns 78° left and we can code this as an *L*-transition. From segment 2 to 3, involves a change of 8° to the right and we code this *S*. The complete sequence is *LSSRSSLRSS*, and we can summarize this as a matrix of transitions from one 'state' to the next in Table 4.3. Tables of this type can be analysed in a number of ways, in order to throw light on the nature of the sinuosity. A more complex alternative, beyond the scope of this book, is to use spectral analysis as pioneered in geomorphology by Speight (1965) (see also Rayner, 1971).

Let us now suppose that instead of describing a single path, we wish to examine the properties of a number of paths across a study area, as illustrated in Figure 4.4(a). Real-world examples might include individual journeys to work, the trajectories of particles in the atmosphere, and so on. A number of measures are possible, of which the simplest is the average path length $\bar{l}$ calculated as

$$\bar{l} = \Sigma l_i / n$$

Table 4.3 Angular transitions for downpath changes in direction (Figure 4.3(d))

	'To'		
'From'	*>45° left*	*straight*	*>45° right*
45° left	0	1	1
straight	1	3	1
45° right	0	2	0

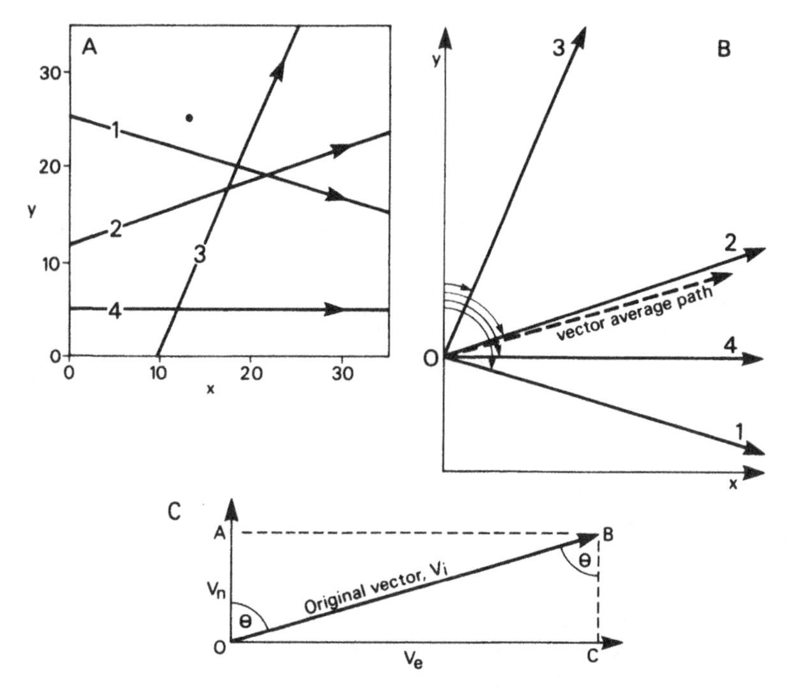

Figure 4.4 a, b and c Paths as vectors.

where n = number of paths. In Figure 4.4(a), the paths have lengths of 36.40, 36.84, 38.08 and 35.00, giving an average of 36.58 units. The spread of lengths around this average can be described by their standard deviation s_l. In the example, this is found as follows:

$$s_l = \left[\sum (l_i - \bar{l})^2 / n \right]^{0.5}$$
$$= (4.8464/4)^{0.5} = 1.10073 \text{ units}$$

A second pair of measures characterize the extent to which paths cover the area in which they lie. These are the *path density:*

$$D = \Sigma l_i / a$$

where a = area, and its reciprocal, the *constant of path maintenance:*

$$C = a / \Sigma l_i$$

Path density has units of a length divided by an area, so its dimension is $LL^{-2} = L^{-1}$. It can be calculated for any line pattern, but has most application in geomorphology as the drainage density, defined as the length of stream channel per unit of drainage basin area. This varies according to factors such as the lithology, rainfall regime and vegetation. The constant of path maintenance can be given a physical interpretation as the length needed to maintain one unit of path length, and has dimension $L^2L^{-1} = L$. In the example, D is $146.32/35^2 = 0.1194$ per unit length, and the constant of path maintenance is 8.372.

Related to these two measures are the *path frequency* F, given as the number of paths per unit area irrespective of their length, and a useful ratio F/D^2, which evaluates the completeness with which a path system fills an area for a given number of paths. This final ratio is particularly useful: as can be seen by examining the dimensions of F and D^2, it has the advantage of being dimensionless. The concept of *path spacing* is inherent in most of the above measures. This has been examined by Dacey (1967), who proposed using measurements made on a number of randomly located traverses across the study area.

A second important property of the individual path is its *direction.* Conventionally, directions are measured as dimensionless quantities called angles, relative to some understood reference, usually but not necessarily north, and these are measured clockwise from this as in Figure 4.3(a). Angular measure gives difficulties not present in ordinary linear measure, which result because values repeat every 360°. The difference between the directions, or bearings, 1° and 359° is not 358°, but only 2° as we pass north. Sometimes this periodic nature of angular data is not a problem, but in most applications we need to take special note of it. Suppose that, for example, we wish to record the angle through which the minute hand of a clock passes in a day. Obviously, this is not 360° : in 24 hours, the hand will sweep this 24 times, giving an angle of 8640°! An alternative measurement unit of angular measure is the *radian,* defined so that there are 2π radians to a circle, and one radian is 57.296°. In a day, the minute hand of the clock sweeps $24 \times 2\pi = 150.8$ radians.

The periodic repetition of angular values is of most concern when we try to find an average path direction. The solution is to treat each path length and direction as a *vector* quantity, having both magnitude (length) and direction as in Figure 4.4(b) and (c). Suppose that, as in the figure, we have four paths of lengths $v_1 \ldots v_4$ and directions $\theta_1, \ldots, \theta_4$. These may be represented as a bundle of vectors from a common origin, as shown in Figure 4.4(b), and we can use vector addition to add them and obtain their *resultant.* This can be measured graphically or calculated with the help of a little trigonometry. In Figure 4.4(c) it can be seen that each individual

Table 4.4 Looking up Trigonometric Ratios

		sign (plus or minus) of result		
Angle, θ	*Look up*	*cosine*	*sine*	*tangent*
0–90	θ	+	+	+
90–180	$180 - \theta$	−	+	−
180–270	$\theta - 180$	−	−	+
270–360	$360 - \theta$	+	−	−
		(*y* slope)	(*x* slope)	(see Chapter 6)

vector can be 'resolved' into two perpendicular components, one expressing its 'northerliness' v_n, the other its 'easterliness' v_e. To find v_n, examine the triangle AOB. The cosine of the vector direction θ_i is given by

$$\cos(\theta_i) = OA/OB$$
$$= v_n/v_i$$

Hence:
$$v_n = v_i\cos(\theta_i)$$

Similarly, the 'easterliness' v_e can be found from

$$\sin(\theta_i) = OC/OB$$
$$= v_e/v_i$$
$$v_e = v_i\sin(\theta_i)$$

All we need do to resolve each vector into perpendicular components, is to multiply the original magnitude by the cosine and sine of the direction. A practical difficulty for those unaccustomed to trigonometric ratios is that published tables list values of sine and cosine only for angles in the range 0°–90°, yet we often need to look up values, such as sin(200°), which are outside this range. Table 4.4 gives the angular value that should be looked up in these circumstances and whether the result is positive or negative. To find sin(200°), look up 200° − 180° = 20°, which has a sine of 0.3420, then give this a negative value, sin(200°) = −0.3420. This resolution into components v_n and v_e is carried out for all paths of interest, as shown in Table 4.5, and the summations of both the northerly and easterly components give a measure of the 'northerliness' and 'easterliness' of the resultant, which can be translated back into a bearing by noting that the tangent of the resultant's direction is

$$\tan\theta_R = \sum v_i\sin\theta_i \Big/ \sum v_i\cos\theta_i$$

Table 4.5 Vector summation, using the paths in Figure 4.4

No.	*Origin*		*End*		*Distance*	*Bearing*	*Resolution*			
	x	y	x	y	v_i	θ^o	$\cos\theta$	$\sin\theta$	$v_i\cos\theta$	$v_i\sin\theta$
1	0	25	35	15	36.40	105	−0.2588	0.9659	−9.420	35.159
2	0	12	35	23.5	36.84	73	0.2924	0.9563	10.772	35.230
3	10	0	25	35	38.08	23	0.9205	0.3907	35.053	14.878
4	0	5	35	5	35.00	90	0.0000	1.0000	0.000	35.000
					146.32		0.9541	3.3129	36.404	120.267

In the example $\Sigma v_i \sin\theta_i$ = 120.27 and $\Sigma v_i \cos\theta_i$ = 36.40, so that

$$\tan\theta_R = 120.27/36.40 = 3.304$$

Referring to tables, this corresponds to the angle 73.25°. In doing this, care should be taken to ensure that this angle is in the correct quadrant according to the signs of the sums of the sine and cosine terms: simply work backwards through Table 4.4. The *vector magnitude* of this resultant can be found from

$$v_r = \left[\left(\sum v_i\cos\theta_i\right)^2 + \left(\sum v_i\sin\theta_i\right)^2\right]^{0.5}$$
$$= [(36.40)^2 + (120.27)^2]^{0.5} = 125.66$$

and the *average vector magnitude* v_r/n = 125.66/4 = 31.41 distance units along the bearing 73.25°, as has been drawn in Figure 4.4(b). It should further be noted that vector addition incorporates directional effects, so that four equal length paths at 90° to one another would have a resultant vector of zero. Vector addition is a useful way of characterizing the net flow of goods, people, or water, but to date very little work has been done in this field and the method finds most application in analysing till fabrics in geomorphology (Pincus, 1956; see also the review by Mardia, 1972).

Trees

The second type of pattern that line data can produce is the branching tree, connecting junctions or nodes but in which no closed loops or circuits are present. Treelike patterns are very common on maps and there has been considerable work on their characteristics, principally by geomorphologists interested in river networks (Werritty, 1972; Gardiner, 1975; Gardiner and Park, 1978), but Haggett (1967) has shown that road nets around a central place can also be broken down into trees.

Although there is a long history of early attempts, most analysis of tree networks owes its inspiration to an article published

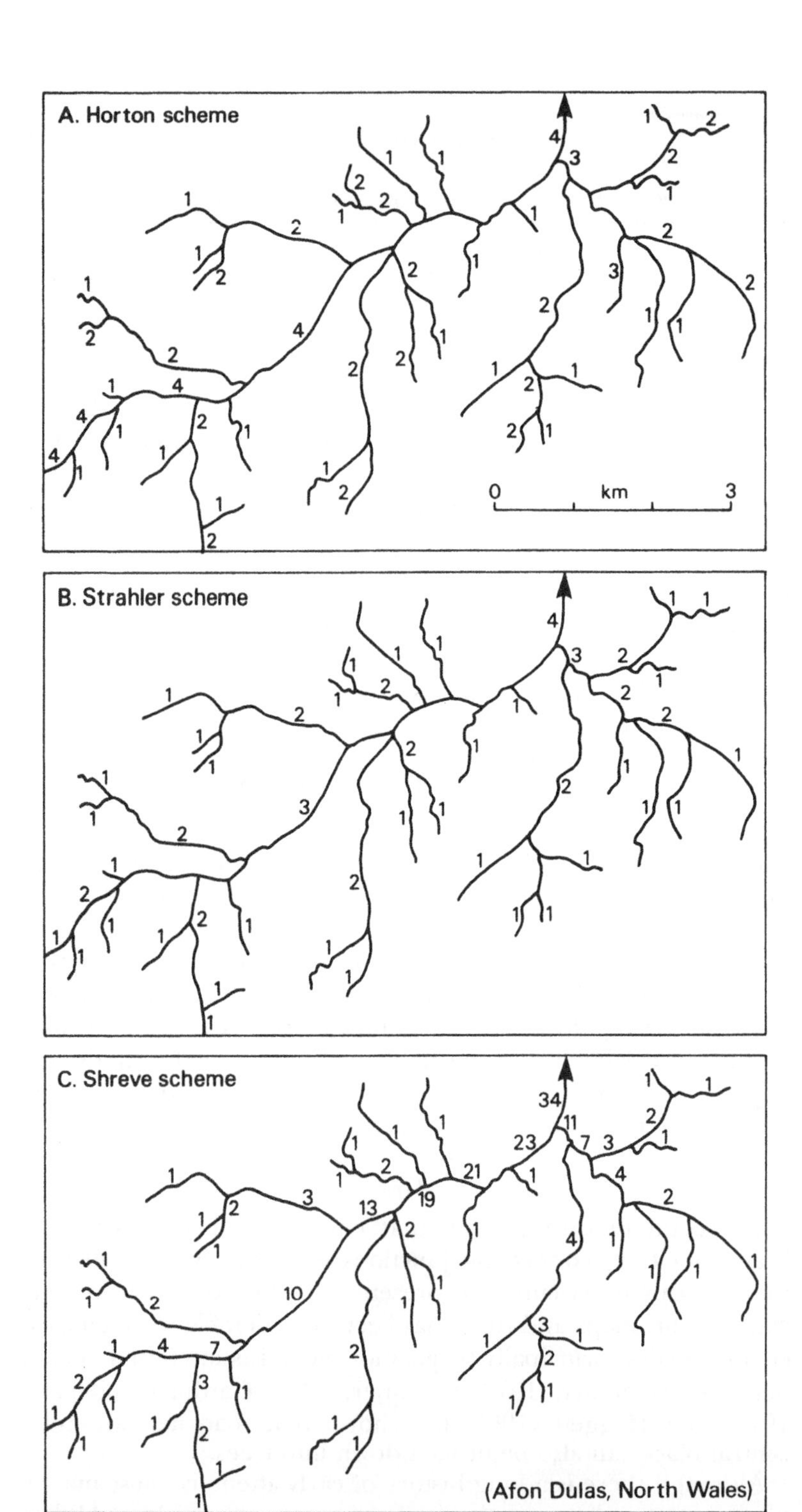

Figure 4.5 a, b and c Methods of ordering paths.

posthumously by an American civil engineer, R. E. Horton, in 1945, proposing a method based on the *ordering* of paths. This is illustrated in Figure 4.5(a), which is actually the drainage network of the Afon (River) Dulas in North Wales traced from a 1 : 50,000 topographic map. There are two steps in the method. First, all fingertip tributaries are given a provisional order of 1, and 2nd, 3rd, 4th, etc. order streams are defined by increasing the order each time two streams of equal order meet. A second order path is made by joining two first orders, and so on, noting that a stream can receive tributaries of lower order without any change in its order. After all the paths in the network are allocated provisional orders, the second step is to work back up the network reclassifying at each junction by continuing the higher-order stream headwards, beginning with the main stream. At any one junction, the main stream is selected as the one having its entrance angle most nearly in line with the main trunk. In practice, this Horton scheme is not easy to adopt and contains an element of subjectivity in the reclassification. It was modified by Strahler (1952), who simply omitted the second step, as shown in Figure 4.5(b). This Strahler scheme has been the most widely used, but like Horton's original, it has the disadvantage that 'extra' streams, such as a first-order stream entering a fourth order, do not affect the total order of the whole network. An alternative, due to Shreve (1966) and illustrated in Figure 4.5(c), gets over this problem by giving each stream a magnitude which is simply the total number of first-order streams headward of it. The final basin order is, thus, an interval-scaled value. Notice that all three methods take no account of stream length, shape, or direction, being concerned only with their interconnections. Such analysis is called *topological.*

Stream-ordering allows objective description of networks, and forms the basis of comparisons and contrasts between differing tree networks. Horton himself proposed a number of regularities in drainage-network composition of which his *law of stream numbers* is the best-known:

> the numbers of streams of different order in a given drainage basin tend closely to approximate an inverse geometric series in which the first term is unity and the ratio is the bifurcation ratio.

Subsequently, workers such as Milton (1966) have shown that this law is maintained in virtually all natural branching systems, including trees and blood veins, and Haggett (1967) refers to it as a

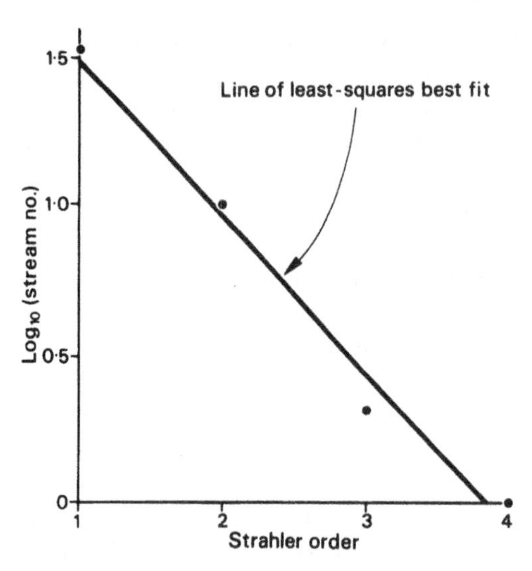

Figure 4.6 The law of path numbers.

very general *law of path numbers*. The law is illustrated in Figure 4.6, which plots the common logarithm of the number of streams in each Strahler order in the Afon Dulas network against the order itself. There are thirty-four first-, ten second-, two third- and one fourth-order streams, so that if the law is maintained the logarithms of these numbers will fall on a straight line. As can be seen, this is roughly the case and a least-squares line of best fit can be inserted as indicated (see Table 4.6).

Because of its almost universal applicability, the fact that tree networks all give straight-line plots of this nature is not of any great practical significance and, indeed, its theoretical justification does not seem very revealing (Werritty, 1972). What is important is that different tree networks generate lines of different slope, as evaluated by the *bifurcation ratio*.

This may be found in any of four ways, each yielding a different result:

(1) As the arithmetic average of the individual ratios of the number of streams in an order to that of the next highest order. From first to second order in the Afon Dulas, this is $34/10 = 3.4$; from second to third order, it is $10/2 = 5$; and from third to fourth order, it is $2/1 = 2$, giving a basin average bifurcation ratio

$$= (3.4 + 5 + 2)/3 = 3.46$$

Table 4.6 Least-squares line of best fit: law of path numbers

Order x	x^2	*Stream number* R	y log(R)	$x\,y$
1	1	34	1.5315	1.5315
2	4	10	1.0000	2.0000
3	9	2	0.3010	0.9030
4	16	1	0.0000	0.0000
10	30	47	2.8315	4.4345

For the straight line, $\log(R) = mx + c$

$$m = \frac{\sum xy - n\overline{xy}}{\sum x^2 - n\overline{x}^2} = -0.53$$

$$c = \overline{y} - m\overline{x} = 2.03$$

Note: For further details of this computation, see Hammond, R. and McCullagh, P. S., 1978, *Quantitative Techniques in Geography*, pp. 101–3.

(2) As a weighted arithmetic average, in which each individual ratio is weighted by the number of streams in the lower order, for example,

$$= \frac{(3.4 \times 34) + (5 \times 10) + (2 \times 2)}{(34 + 10 + 2)} = 3.69$$

(3) As the geometric mean of the individual ratios, that is

$$= \text{antilog}\left[\frac{\log(3.4) + \log(5) + \log(2)}{3}\right] = 3.24$$

(4) Finally, Maxwell (1955) suggested the antilog of the gradient constant, *m*, in the least-squares best fit straight line

$$= \text{antilog } (0.53) = 3.38$$

Whatever method is used, bifurcation ratios must have a minimum value of 2, and natural drainage networks seem to have values from 3 to 5, varying according to mean rainfall, vegetation, geology, and so on.

Many other measures can be developed to describe tree networks. Examples include the frequency distributions of link lengths and entrance angles, as well as the analysis by point pattern methods of the pattern of sources and junctions (Jones, 1978). In transport studies the origins and destinations of flows – the nodes – are more significant than in drainage studies, and such a net can be thought of as a point pattern of *n* nodes which we want to connect in some way. If the aim is to minimize the total length of path involved, a *minimum tree* can be produced and used as a standard against which to measure real-world routes. This is constructed by joining each node to its nearest neighbour, repeating the process until every point is connected (see Getis and Boots, 1978, p. 99).

Circuits

The final type of line pattern is the circuit, in which paths may form closed, connecting loops. The property of *connection* most characterizes this type of network, and its measurement has been considered by a number of workers, particularly in transport studies. Transport networks of road, rail, canal and airways are good examples of circuits. The description of connection is best achieved using the idea of a topological graph (see Haggett and Chorley, 1969; Tinkler, 1977) on which the original network, such as the major roads of the Isle of Skye shown in (a) in Figure 4.7, is reduced to the simplified form shown in (b). This graph is made up of *points* (vertices, nodes) connected by paths (lines, links, edges, connections, relations), but as the figure shows, there is no concern for length, sinuosity, or even direction. The paths themselves may, however, be directed, indicating a one-way flow and giving a directed graph or *digraph.* They may also be given *z*-values according to their flow, nature, or capacity, giving a *valued graph.* The graph drawn in Figure 4.7(b) is also *planar,* that is drawn on a flat surface so that no paths intersect and all paths end or meet at points. A *non-planar graph* is one that cannot be so drawn. This idea of a graph is very general: in geography, graphs have been used to study obvious problems in transportation and glaciated valleys, but they also find use in a wide range of other problems including industrial linkage, periodic markets, and so on.

The graph of a circuit network greatly simplifies the structure of the network, but it soon becomes unwieldy when dealing with large numbers of interconnected points. Such large networks are best

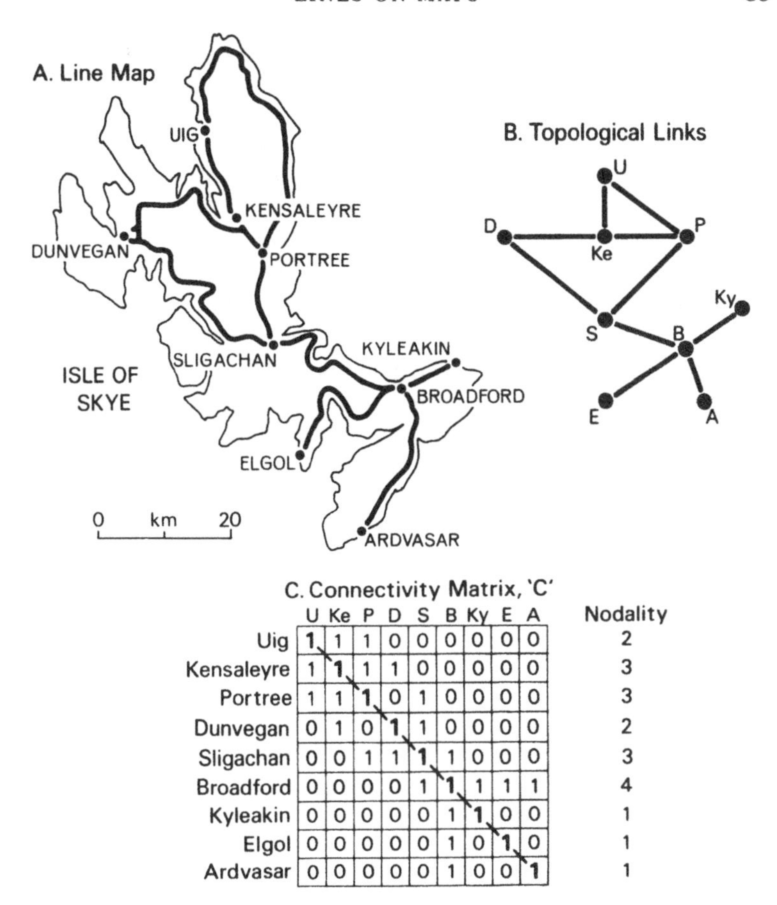

	U	Ke	P	D	S	B	Ky	E	A	Nodality
Uig	**1**	1	1	0	0	0	0	0	0	2
Kensaleyre	1	**1**	1	1	0	0	0	0	0	3
Portree	1	1	**1**	0	1	0	0	0	0	3
Dunvegan	0	1	0	**1**	1	0	0	0	0	2
Sligachan	0	0	1	1	**1**	1	0	0	0	3
Broadford	0	0	0	0	1	**1**	1	1	1	4
Kyleakin	0	0	0	0	0	1	**1**	0	0	1
Elgol	0	0	0	0	0	1	0	**1**	0	1
Ardvasar	0	0	0	0	0	1	0	0	**1**	1

Figure 4.7 a, b and c Reducing a line map to a connectivity matrix.

represented as *connectivity* or *adjacency matrices,* denoted **C,** of the type shown in Figure 4.7(c). To construct such a matrix, the points of the graph are labelled or numbered and the labels used to identify rows and columns of the matrix. If two points are connected, this is recorded by placing a non-zero number at the intersecton of the relevant row and column. If they are not connected, a zero is inserted instead. In the example, Uig is connected to both Kensaleyre and Portree by a simple link we have coded 1, but not to anywhere else. Notice three things about the completed matrix. First, the entries are binary attributes (0/1) measuring the presence or absence of a link, but we could equally well have inserted some ordinal or interval scaled numbers to express the nature of the links. Second, the principal diagonal of the matrix, from top left to bottom right, is coded 1, implying that all places are connected to themselves. In much of the literature, it is claimed that this decision is of no consequence, but as will be seen when we come to power this matrix, using 1s avoids an undesirable side-effect that is noted by Tinkler (1977, p. 24) and analysed by Garner and Street (1978). Third, our matrix is also symmetric about this principal diagonal, that is the lower left triangle of numbers is a mirror-image of the upper right. This results from the use of undirected paths; had we used a matrix to represent a digraph, this symmetry would have been lost.

Given the graph of a network, or its connectivity matrix **C**, there are a great many measures that can be devised, some involving relatively simple combinations of the numbers of points and paths, others rather more complex matrix theory. The simple measures are based on three numbers, which in the usual notation are symbolized as follows:

e = the number of paths in the network;
v = the number of points;
g = the number of separate, non-connecting subgraphs.

In the example $e = 10$, $v = 9$ and $g = 1$. In a planar graph, the number of basic circuits is called μ, the *cyclomatic number*, defined as

$$\mu = e - v + g$$

This will obviously increase as the network becomes more connected to a maximum value determined by the number of points. For the Skye network

$$\mu = 10 - 9 + 1 = 2$$

To compare networks of differing size, it is more useful to relate the cyclomatic number to the maximum number of circuits, as in the *alpha index:*

$$\alpha = \mu/(2v - 5)$$

which ranges from 0 (non-connected paths) to 1 for a fully connected net. The Skye network has a value of 0.154, indicating that it has relatively few circuits. The *beta index* is simply

$$\beta = e/v$$

but like the cyclomatic number, it varies according to the number of points and as such is not very useful in comparisons. For Skye it is 10/9 = 1.1. Again, however, we can express it as a ratio of the number of paths to the maximum number of paths given a specified number of points. This gives the *gamma index:*

$$\gamma = e/[3(v\text{-}2)]$$

which, like alpha, will vary from 0 to 1. The Skye network has

$$\gamma = 10/3(9 - 2) = 0.48$$

If, instead of counting paths and points, we examine the connectivity matrix **C**, then further indices can be proposed. The *nodality* of each node is the number of paths incident at that point, and the *mean nodality* for the network is simply the average of these values. As Figure 4.7(c) shows, nodality is found simply by summing each row of the **C** matrix less one for the diagonal. For Skye, values vary from 1 to 4, with Broadford the most connected place, and there is a low mean nodality of 2.2. To investigate variations around this mean, it is also useful to assemble and compare frequency distributions of nodality, as in the next section, pp. 103–6.

Our final series of circuit measures depend on rather more arithmetical manipulation of **C**. Just as ordinary numbers can be multiplied together, so it is possible to multiply whole matrices, and it turns out that successive multiplication of **C** by itself, called for obvious reasons *powering* **C**, yields valuable structural information about the network. The only problems are how to power a matrix and the amount of labour involved. Matrix multiplication of a small square matrix is actually simple: all one needs to do to find each position, or *element,* of the result is to multiply each row of the matrix by each column, taking the sum as the result. For a simple two-row, two-column matrix this would be

$$\mathbf{M}^2 = \mathbf{M} \times \mathbf{M} = \begin{pmatrix} a & b \\ c & d \end{pmatrix} \times \begin{pmatrix} a & b \\ c & d \end{pmatrix} = \begin{pmatrix} aa + bc & ab + bd \\ ca + dc & cb + dd \end{pmatrix}$$

For example, if $a = 1$, $b = 2$, $c = 3$ and $d = 4$, we have

$$\mathbf{M}^2 = \begin{pmatrix} 1 & 2 \\ 3 & 4 \end{pmatrix} \times \begin{pmatrix} 1 & 2 \\ 3 & 4 \end{pmatrix} = \begin{pmatrix} 1 + 6 & 2 + 8 \\ 3 + 12 & 6 + 16 \end{pmatrix} = \begin{pmatrix} 7 & 10 \\ 15 & 22 \end{pmatrix}$$

For a small matrix this is all very easy, but as the number of rows and columns increases, so does the labour involved. For a large matrix, like the **C** derived for the Skye network, it is better to isolate each stage of the calculation, or use a computer. In the Skye case, the first element of the product $\mathbf{C}^2$ (i.e. Uig to Uig!) is found as:

Row		*Column*	
1	×	1	= 1
1	×	1	= 1
1	×	1	= 1
0	×	0	= 0
0	×	0	= 0
0	×	0	= 0
0	×	0	= 0
0	×	0	= 0
0	×	0	= 0
			3

For the second element, Uig to Kensaleyre the *row* is the same, but this time we use the second column:

Row		*Column*	
1	×	1	= 1
1	×	1	= 1
1	×	1	= 1
0	×	1	= 0
0	×	0	= 0
0	×	0	= 0
0	×	0	= 0
0	×	0	= 0
0	×	0	= 0
			3

and so on. The completed calculation for the second power of **C** is:

$$\mathbf{C}^2 = \mathbf{C} \times \mathbf{C} = \begin{pmatrix} 3 & 3 & 3 & 1 & 1 & 0 & 0 & 0 & 0 \\ 3 & 4 & 3 & 2 & 2 & 0 & 0 & 0 & 0 \\ 3 & 3 & 4 & 2 & 2 & 1 & 0 & 0 & 0 \\ 1 & 2 & 2 & 3 & 2 & 1 & 0 & 0 & 0 \\ 1 & 2 & 2 & 2 & 4 & 2 & 1 & 1 & 1 \\ 0 & 0 & 1 & 1 & 2 & 5 & 2 & 2 & 2 \\ 0 & 0 & 0 & 0 & 1 & 2 & 2 & 1 & 1 \\ 0 & 0 & 0 & 0 & 1 & 2 & 1 & 2 & 1 \\ 0 & 0 & 0 & 0 & 1 & 2 & 1 & 1 & 2 \end{pmatrix}$$

Next, multiply this by **C** again to find $\mathbf{C}^3$ as:

$$\begin{pmatrix} 9 & 10 & 10 & 5 & 5 & 1 & 0 & 0 & 0 \\ 10 & 12 & 12 & 8 & 7 & 2 & 0 & 0 & 0 \\ 10 & 12 & 12 & 7 & 9 & 3 & 1 & 1 & 1 \\ 5 & 8 & 7 & 7 & 8 & 3 & 1 & 1 & 1 \\ 5 & 7 & 9 & 8 & 10 & 9 & 3 & 3 & 3 \\ 1 & 2 & 3 & 3 & 9 & 13 & 7 & 7 & 7 \\ 0 & 0 & 1 & 1 & 3 & 7 & 4 & 3 & 3 \\ 0 & 0 & 1 & 1 & 3 & 7 & 3 & 4 & 3 \\ 0 & 0 & 1 & 1 & 3 & 7 & 3 & 3 & 4 \end{pmatrix}$$

Finally, repeat to find $\mathbf{C}^4$ as:

$$\begin{pmatrix} 29 & 34 & 34 & 20 & 21 & 6 & 1 & 1 & 1 \\ 34 & 42 & 41 & 27 & 29 & 9 & 2 & 2 & 2 \\ 34 & 41 & 43 & 28 & 31 & 15 & 4 & 4 & 4 \\ 20 & 27 & 28 & 23 & 25 & 14 & 4 & 4 & 4 \\ 21 & 29 & 31 & 25 & 36 & 28 & 12 & 12 & 12 \\ 6 & 9 & 15 & 14 & 28 & 43 & 20 & 20 & 20 \\ 1 & 2 & 4 & 4 & 12 & 20 & 11 & 10 & 10 \\ 1 & 2 & 4 & 4 & 12 & 20 & 10 & 11 & 10 \\ 1 & 2 & 4 & 4 & 12 & 20 & 10 & 10 & 11 \end{pmatrix}$$

At this step it will be observed that all the zeros in the original **C** have been lost, and this is as far as it is necessary to go in the powering process.

What useful results have been obtained? The elements of the powers of the connectivity matrix contain a great deal of information about the network, but the interpretation of these elements differs according to whether the original matrix has zeros or ones in the principal diagonal. Starting with zeros, gives elements of each nth power that are the number of routes of *exactly n* links between each pair of points. For the Skye network it is possible to join any two places by a chain of a maximum of four links, this being called the *graph diameter*. In much of the literature, it is claimed that a connectivity matrix powered by as many times as the graph diameter will contain no zero entries, but in fact, this is not generally true. The key word in our description is *exactly n* links: it is possible for two places to be connected by, say, a chain of three links but have no possible four-link connection.

If 1s are used in the principal diagonal, the elements of each nth power give the number of connections of *at most n* steps, that is it has added a memory of steps of lesser numbers of links. Garner and Street (1978) show that this past memory (at least in the off-diagonal elements) can be broken down into its constituents to give exactly the same information on exactly n-step routes that would be obtained using zeros. The off-diagonal elements of a connectivity matrix to the power n are the sum of the number of links of exactly n steps plus a weight times the number of routes of exactly $n - 1$ steps, plus a second weight times the number of exactly $n - 2$ steps, and so on. Irrespective of the original matrix, these weights are constant and are listed in Table 4.7 for up to $n = 8$.

For example, examine the entries in our powered matrix for links between Sligachan (row 5) and Uig (column 1). There is no direct link, so that the entry in $\mathbf{C}^1$ is zero. In $\mathbf{C}^2$, it is 1; in $\mathbf{C}^3$, 5; and in $\mathbf{C}^4$, 21, and these represent the total number of *at most n* steps. We can recover the exact numbers as follows. At the second power $n = 2$, our value of 1 is made up as:

$$
\begin{aligned}
1 = 1 &\times \text{exact number of two-step links} = (1 \times ?) = ?\\
+ 2 &\times \text{exact number of one-step links} = (2 \times 0) = \underline{0}\\
&\qquad\qquad\qquad\qquad\qquad\qquad\qquad\qquad\qquad\quad 1
\end{aligned}
$$

There must be just one link of *exactly* two steps, which inspection of Figure 4.7(a) shows is the route round the north of the island to Portree. At the third power our entry of 5 is made up as:

$$
\begin{aligned}
5 = 1 &\times \text{exact number of three-step links} = 1 \times ? = ?\\
+ 3 &\times \text{exact number of two-step links} = 3 \times 1 = 3\\
+ 3 &\times \text{exact number of one-step links} = 3 \times 0 = \underline{0}\\
&\qquad\qquad\qquad\qquad\qquad\qquad\qquad\qquad\quad 5
\end{aligned}
$$

Hence, there are just two links of three steps. Finally, in the final matrix $\mathbf{C}^4$, the entry is 21 made up as:

$$
\begin{aligned}
21 = 1 &\times \text{exact number of four-step links} = 1 \times ? = ?\\
+ 4 &\times \text{exact number of three-step links} = 4 \times 2 = 8\\
+ 6 &\times \text{exact number of two-step links} = 6 \times 1 = 6\\
+ 4 &\times \text{exact number of one-step links} = 4 \times 0 = \underline{0}\\
&\qquad\qquad\qquad\qquad\qquad\qquad\qquad\qquad\quad 21
\end{aligned}
$$

Table 4.7 Weights for the off-diagonal elements of a powered connectivity matrix, starting with 1s in the diagonal (off-diagonal elements made up of the weight below multiplied by the exact number of steps)

Power	*No. of steps*							
n	*n*	*n* - 1	*n* - 2	*n* - 3	*n* - 4	*n* - 5	*n* - 6	*n* - 7
1	1							
2	1	2						
3	1	3	3					
4	1	4	6	4				
5	1	5	10	10	5			
6	1	6	15	20	15	6		
7	1	7	21	35	35	21	7	
8	1	8	28	56	70	56	28	8

We conclude that there must be seven links of exactly four steps. The *diagonal entries* of our final solution matrix have been abstracted in the second column of Table 4.8. Their interpretation in terms of numbers of routes is complex, but they do give a good relative index of the connectivity of each point in the network. As the table indicates, we can also find the mean of these values as a measure of the overall connectivity of this particular network.

A further matrix that can be derived from **C** is that of the shortest paths, again measured topologically as the number of intervening links. For Skye this is:

	U	K	P	D	S	B	K	E	A	*Sum of off-diagonals*
U	1	1	1	2	2	3	4	4	4	21
K	1	1	1	1	2	3	4	4	4	20
P	1	1	1	2	1	2	3	3	3	16
D	2	1	2	1	1	2	3	3	3	17
S	2	2	1	1	1	1	2	2	2	13
B	3	3	2	2	1	1	1	1	1	14
K	4	4	3	3	2	1	1	2	2	21
E	4	4	3	3	2	1	2	1	2	21
A	4	4	3	3	2	1	2	2	1	21

It can be obtained either directly, by inspection of the original graph, or as a byproduct of the powering of **C**, since each element in it is the power at which **C** becomes non-zero. A number of indices can be obtained from this matrix. The maximum number on each row, indicating the number of steps needed to reach the

Table 4.8 Connectivity measures based on the connection and distance matrices for the Skye main-road network

	From $\mathbf{C}^1$	*From* $\mathbf{C}^4$	*From distance matrix*		
Point	*Nodality*	*Connectivity*	*König no.*	*Access sum*	*Average distance*
Uig	2	29	4	21	2.63
Kensaleyre	3	42	4	20	2.50
Portree	3	43	3	16	2.00
Dunvegan	2	23	3	17	2.13
Sligachan	3	36	2	13	1.63
Broadford	4	43	3	14	1.75
Kyleakin	1	11	4	21	2.63
Elgol	1	11	4	21	2.63
Ardvasar	1	11	4	21	2.63
	20	249		164	

Mean nodality = 20/9 = 2.22
Graph dispersion = 164
Mean connectivity = 249/9 = 27.67

topologically most distant point, is called the *König number,* and the maximum of these is our graph diameter. As Table 4.8 shows, Sligachan has the lowest König number. Summing the off-diagonal elements of each row (see Table 4.8) gives an *access sum* whose average over all points, after division by $n - 1$, is a measure of the average distance, again in topological steps, to other places. The original summation, over all places, is called the *graph dispersion.* Research has shown that in practice dispersion values by themselves are not very useful; what has proved more worthwhile is the analysis of the frequency distributions of individual values (see Ord, 1967; James *et al.*, 1970).

The great wealth of indices of connection for circuit networks introduced in this section (and there are more!) lays the quantifier open to charges of 'playing with numbers for numbers' sake'. Quite clearly, the single geographical concept of connection can be measured in any of a number of ways, each with a slightly different interpretation. As Taylor (1977, p. 63) points out, this 'can be a dangerous, or, at least, a misleading luxury', forcing any researcher to be very careful to ensure that the most useful index is used to indicate what is intended. In such circumstances, these measures can provide a useful starting-point in investigations, such as the effect of removing or adding links on overall network accessibility, in comparing more than one network, or in attempts to isolate the circuit-generating processes that created it.

ANALYSING LINES ON MAPS

In Chapter 3 (pp. 48–62) methods of analysing some of the various measures of point patterns for evidence of a non-random generating process were examined. The important concepts that were introduced included, first, the idea of an observed map as a realization of a stochastic process in which the location of each point symbol was controlled by random selection from an underlying, fixed probability distribution. Second, the specific process we used as a standard against which to evaluate patterns was the independent random. Third, a statistical significance test was carried out to ascertain the probability of a particular statistic occurring by chance with a true null hypothesis of independent randomness. In this section, we shall attempt to apply similar ideas to the various measures of path length, direction and connection, beginning with an analysis of simple paths, going on to examine trees, and ending with an analysis of circuits.

Paths across and in areas

Consider a blank area such as an open park or plaza to be crossed by shoppers and across which no fixed paths exist. The analogous process to the independent random location of a point is to select a location on the perimeter of the area allowing each point an equal and independent chance of being selected, then to draw a path in a random direction from it until it reaches the perimeter (or to select randomly a second point on the perimeter and join up the points). Next proceed to produce a series of these lines, so that the pattern they make is one realization of the random process. We can summarize our results by the application of any of the methods given on pp. 72–9, but for simplicity we shall examine only the frequency distribution of path lengths.

What values would we expect, in the long run, from this independent random process? Although the general principles are the same, deducing the expected frequencies of path lengths given an independent random process is rather more difficult than it was

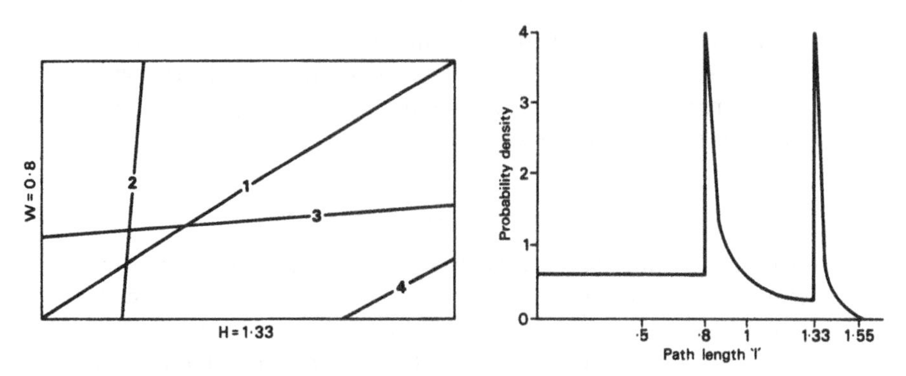

Figure 4.8 Path lengths across a rectangular area.

for point patterns. There are three reasons for this. First, it will be remembered that the frequency distribution of quadrat counts was discrete, that is it could only have steps of whole numbers corresponding to cell counts with $m = 1, 2, 3, \ldots, n$ points in them. Path lengths can take on any value, so the distribution involved is a *continuous* probability density function. In practice, this makes the mathematics a little more difficult. Second, a moment with a pencil and paper will indicate that, because they are constrained by the perimeter of the area, path lengths strongly depend on the shape of the area they cross. Third, mathematical statisticians have paid less attention to path-generating process than they have to point ones. An exception is the work of Horowitz (1965), described by Getis and Boots (1978), and what follows is based very largely on his paper.

Starting from the independent random assumptions already outlined, Horowitz was able to derive the probabilities of paths of given length for five basic shapes, the square, rectangle, circle, cube and sphere. His results for a rectangular area of side 1.33×0.8 units of length are shown in Figure 4.8. There are several points to note: the probability associated with any exact path length in a continuous probability distribution is very small, thus what is plotted in (b) in the diagram is the probability density, that is, the probability per unit change in length. In addition, this probability density can be seen to be very strongly influenced by the area shape. The longest path possible across our rectangle, shown as path 1 in (a) in the figure has a length of 1.55 units, and this sets an upper limit on the distribution. Moreover, the independent random process will tend to give higher probabilities of lines of lengths close to 0.8 and 1.33, for example, paths 2 and 3 in (a), than shorter paths of the type shown as path 4. The result is a very markedly bimodal probability density function. For a square area, with equal width and height, the function is unimodal.

We can use the probabilities predicted by Horowitz's model to construct a significance test against any observed distribution. For simplicity, this analysis will be restricted to a *circular* area of radius equal to 1 unit, but if the observed lengths of path across any circular area are standardized by dividing by its radius, then the expected values quoted will still be valid. The first two columns in Table 4.9 give the expected proportions falling into the specified path lengths, up to the maximum possible across a diameter of the circle. These have been cumulated to give column 3. In order to test the model, and to demonstrate how a significance test might be applied, column 4 gives the observed results of an experiment in which 100 paths were drawn across a unit circle according to our independent random rules, and column 5 gives these frequencies as cumulative proportions of the total. Figure 4.9 shows the first ten of these. The simplest test of significance we can use is the Kolmogorov-Smirnov (for a full account, see Siegel, 1956), in which we find the largest absolute difference in cumulative proportions between the observed frequencies (column 5) and expected (column 3). This turns out to be 0.045 which, with 100 observations, is well below the tabulated critical value at the 95 per cent level of 0.136. We conclude that the observed frequencies do not differ significantly from the expected and, since the observed are in actual fact one realization of the random process, this result is not particularly surprising; but it does illustrate how a significance test may be constructed. The remaining columns of the table repeat this analysis for samples of 1000 and 10,000 random lines across a circle (since nobody in his right mind would actually draw 10,000 random lines, the perceptive reader will have guessed that these results came from a computer simulation). Notice that as the simulated sample size increases, the proportions get very close to those predicted by Horowitz's mathematics.

There are a number of very practical situations where the statistical properties of straight-line paths across specific shapes are required, but these seem to occur mostly in physics (gamma-rays across a reactor, sound waves in a room, and so on) rather than geography. A possible application, to pedestrian paths across a circular shopping plaza, is given in Getis and Boots (1978); the

Table 4.9 Probabilities and observed proportions of path lengths across a unit circle

			Simulation results					
	Expected	*Cumulative*	*n = 100*		*n = 1000*		*n = 10,000*	
Path length	*proportion*	*expected*	*Observed*	*Cumulative*	*Observed*	*Cumulative*	*Observed*	*Cumulative*
0–0.20	0.064	0.064	6	0.060	55	0.055	624	0.062
0.21–0.40	0.064	0.128	6	0.120	61	0.116*	642	0.127
0.41–0.60	0.066	0.194	5	0.170	66	0.182	646	0.191
0.61–0.80	0.068	0.262	11	0.280	75	0.257	750	0.266
0.81–1.00	0.071	0.333	5	0.330	76	0.333	755	0.342*
1.01–1.20	0.076	0.409	7	0.400	70	0.403	717	0.413
1.21–1.40	0.084	0.493	8	0.480	82	0.485	845	0.498
1.41–1.60	0.096	0.589	12	0.600	93	0.578	918	0.590
1.61–1.80	0.116	0.705	15	0.750*	122	0.700	1226	0.712
1.81–2.00	0.296	1.000	25	1.000	300	1.000	2877	1.000
			d_{max}	= 0.045		= 0.012		= 0.009
			d_{crit}	= 0.136		= 0.043		= 0.014

* Maximum differences.

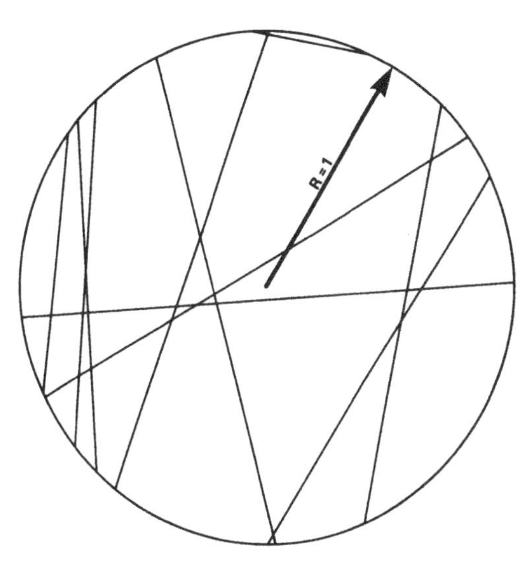

Figure 4.9 Path lengths across a circular area.

reader may well be able to list others. A major difficulty is that few geographical areas of interest have the simple regular shapes that allow mathematical derivation of the probabilities. Instead, it would almost certainly prove necessary to use computer simulation to establish the expected, independent random probabilities appropriate to these more complex shapes.

A related, even more complex, problem that has more applicability in geography is that of establishing the probabilities of all possible distances *within* irregular shapes, rather than simply across the shape as in the Horowitz model. Practical applications would be to the lengths of journeys in cities of various shapes, the distances between the original homes of marriage partners, and so on. Given such data, the temptation is to test the observed distribution of path lengths against some uniform or random standard without taking into account the constraints imposed by the shape of the study area. Yet, as the pioneer paper by Taylor (1971) shows, this shape *does* influence the obtained frequency distribution of path lengths and it is this constrained distribution that ought to be used as a standard. As was suggested above, Taylor found it necessary to use computer simulation rather than mathematical analysis.

The test outlined above enables the investigator to test for an independent random process generating the observed path lengths. Can we devise a similar test for the next fundamental spatial concept, direction? A number of such tests have been devised and used, largely by geologists interested in sediments such as glacial tills in which the orientations of the particles have process implications. A comprehensive review of the field, which is required reading for anyone with more than a passing interest, is the book by Mardia (1972).

Early tests were based on unwrapping circular distributions and applying conventional statistics, but it is better to use a test which makes explicit use of the periodic property of directional measures, such as that devised a century ago by Lord Rayleigh (see Curray, 1956; Durand and Greenwood, 1958). Given n observations of path direction θ_i, the first step is to resolve each into its perpendicular components. The procedure is similar to that outlined on pp. 77–9 and used in Table 4.5, except that no multiplication of the direction sines and cosines by the vector magnitude is undertaken; hence:

$$v_n = \sum_{i=1}^{n} \cos\theta_i$$

$$v_e = \sum_{i=1}^{n} \sin\theta_i$$

The resultant vector magnitude is:

$$v_R = (v_n^2 + v_e^2)^{0.5}$$

Rayleigh showed that the quantity $2v_R^2/n$ can be treated as a chi-square value with two degrees of freedom, whose probability, under a null hypothesis of an independent random process, can be found from tables in the normal way.

In the example given in Figure 4.4 and listed in Table 4.5, we have:

$$n = 4$$

$$v_n = \sum \cos\theta_i = 0.9541$$

$$v_e = \sum \sin\theta_i = 3.3129$$

$$v_R = (v_n^2 + v_e^2)^{0.5} = (0.9541^2 + 3.3129^2)^{0.5} = 3.4476$$

Hence $2v_R^2/n$ is 5.942. The tabulated critical chi-square at the 95 per cent confidence level with two degrees of freedom is 5.9915. The observed value is just short of this so that, strictly speaking, we cannot reject the null hypothesis. Numerous other tests may be performed on directional data; a particularly useful one is Kuiper's test, which uses the frequency distribution of directions (see Lewis, 1977, p. 228). As was shown on pp. 55–62 when dealing with point data, a rejection of the null hypothesis indicates that the distribution of path directions is not independent random; some directional bias is present. In till fabric analysis such a bias is indicative of the direction of a glacier flow, but in transport geography it could indicate a directional bias imparted by the pattern of valleys along which the easiest routeways were found, and so on.

Tree networks

We have seen how tree networks may be described using the ordering method. Our task here is to devise a method that will give the probability of any particular observed tree network, given all the alternative networks that could have been formed by random linking of paths subject to the basic constraint that circuit-forming links are not allowed. This statistical approach to tree networks has been developed by geomorphologists interested in drainage patterns, and it is particularly to the work of Shreve (1966) that we owe the link between Horton's original ideas and stochastic process theory.

The underlying idea used is that of *topologically distinct channel networks* (TDCNs), or networks which look different when reduced to a topological graph but have an equal number of stream sources and links. For example, Figure 4.10 shows all the possible TDCNs for a stream network with five sources. The total number of TDCNs for a given number of sources n, given by random joining, can be calculated as:

$$N(n) = \frac{1}{2n - 1}\binom{2n - 1}{n}$$

For five sources we have:

$$N(5) = \frac{1}{10 - 1}\binom{10 - 1}{5} = \left(\frac{1}{9}\right)\frac{9!}{5!4!} = 14$$

which is exactly the number drawn. The number of possible TDCNs rises very rapidly as n is increased. For $n = 6$ it is 42, for $n = 9$ it is 1430 and, as will be seen later, for $n = 34$ it is 2.1313×10^{17}.

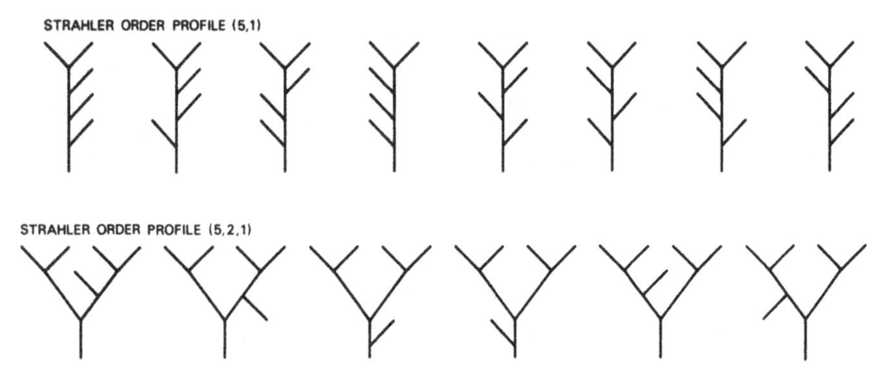

Figure 4.10 Topologically distinct channel networks for magnitude-5 streams.

Such large numbers make testing the significance of any particular arrangement very difficult. A possible solution is to group the TDCNs into classes according to their 'profile' of Strahler stream number counts–after all, this is usually what is measured! For example, the fourteen five-source networks in Figure 4.10 can be grouped into only two profiles of stream counts, designated as $N(5,1)$ and $N(5,2,1)$. As the figure shows, there are eight TDCNs that have five first- and one second-order stream ($N(5,1)=8$) and six that have five first, two second and one third ($N(5,2,1)=6$). Having performed this count, we can now compare any observed order profile with what would be expected by random linkage, but in order to do this, it is necessary to be able to find the numbers of TDCNs in each possible order profile generated by such a process. Shreve (1966, p. 29) gives this as:

$$N(n_1, n_2, \ldots, n_{m-1}, 1) = \prod_{w=1}^{m-1} 2^{(n_w - 2n_{w+1})} \binom{n_w - 2}{n_w - 2n_{w+1}}$$

This fortunately, can be cut down to size fairly easily. First and foremost, the principle is just the same as we have used all along. It is a mathematical expression for the number of TDCNs in each Strahler order profile expected from a random process model. The left-hand side of the equation is simply a shorthand way of writing this in which n_1 is the number of first-order streams, n_2 the second order, and so on until, at the maximum order, m there is just one stream. On the right-hand side, n_w is simply the number of streams in order w, so that n_{w+1} is the number in the next higher order. The symbol Π is a mathematical operator similar to the by now familiar Σ but, instead of indicating the result of a sequence of additions (a sum), it indicates the result of a series of multiplications (a product). The final part is the shorthand for the number of combinations introduced on p. 51.

Putting some numbers into the formula, will make things clearer. Remember that from our sketch (Figure 4.10) we know that a five-source network has only fourteen possible TDCNs, and that these can be grouped into eight of the Strahler order profile (5,1) and six of (5,2,1). The formula does away with the need to draw and count these TDCNs. For example, the number of TDCNs with the profile (5,2,1) is given by substitution into the formula. With $w = 1$, $n_w = 5$ we have:

$$N(5,\ 2,\ 1) = 2^{(5-2\times 2)}\binom{5-2}{5-2\times 2} \times 2^{(2-2\times 1)}\binom{2-2}{2-2}$$

$$= 2^1\binom{3}{1} \times 1 = 2\left(\frac{3!}{1!\ \ 2!}\right) = 6$$

(Note that by definition $2^0 = 1$ and $0! = 1$.) The result is the same as obtained by drawing. Similarly, the number of TDCNs with order profile (5,1) is given by:

$$N(5,\ 1) = 2^{(5-2\times 1)}\binom{5-2}{5-2\times 1} = 2^3 = 8$$

If further explanation is necessary, the Worksheet at the end of this chapter has an exercise involving these calculations for a nine-source network. Complete it and then return to this point.

The value of Shreve's formulae is that they enable us to evaluate the probability of any observed Strahler order profile under a random topology hypothesis analogous to that used in the tests we have already considered in quadrat counts and path lengths. Consider the stream network of the Afon Dulas mapped in Figure 4.5. It has thirty-four sources and a Strahler order profile of (34, 10, 2, 1). If we are prepared to do an amount of arithmetic and deal with some astronomically high numbers, we can find the probability of this particular profile being produced by the random topology model as follows.

First, we find the *total* possible TDCNs from a thirty-four-source network. From the first of Shreve's formulae this is:

$$N(n) = \frac{1}{2n-1}\binom{2n-1}{n} = \frac{1}{67}\binom{67}{34} = \frac{1}{67}\left(\frac{67!}{34!\ \ 33!}\right)$$

There is an immediate difficulty. How can anyone be expected to evaluate 33! (= 33 × 32 ×, . . . ,× 1), let alone 67!? Where small numbers are involved, it is usually possible to cancel out a lot of the multiplications, but with numbers this high, it is best to use Stirling's approximation (see Wilson and Kirkby, 1980, p. 258)

$$n! \approx 2.506628\, n^{(n+0.5)} \exp(-n)$$

(≈ denotes approximately equal to). Using this, we find that there are approximately 2.13126×10^{17} TDCNs from a thirty-four-source network.

Next, we need to find out how many of these TDCNs have the same Strahler order profile as observed, that is (34, 10, 2, 1). This is given by the second Shreve formula as:

$$N(34, 10, 2, 1) = 2^{34-20}\binom{34-2}{34-20} \times 2^{10-4}\binom{10-2}{10-4} \times 2^{2-2}\binom{2-2}{2-2} = 1.39523 \times 10^{16}$$

Finally, we can express these two numbers as a probability:

$$\text{probability (profile 34, 10, 2, 1)} = \frac{\text{number of TDCNs with this profile}}{\text{total number of TDCNs}} = \frac{1.39523 \times 10^{16}}{2.13126 \times 10^{17}} = 0.0655$$

In other words, it is a fairly common profile, likely to be found in about 6.5 per cent of all TDCNs produced by a random model. It is certainly much more likely to occur randomly than the extreme profile (34, 1), which has a probability of 0.0000000326. In fact, this specific result for the Dulas is a very general one. Natural tree networks give Strahler order profiles that are fairly probable realizations of a random model. The geomorphological consequences of this discovery, together with further statistical tests for tree networks and an in-depth review are to be found in Werritty (1972). For a recent dissenting view, see Jones (1978).

Circuit networks

In the past geographers have usually taken the structure of a network expressed as a topological graph as given, attempting to relate its structure to flows along the various paths. In contrast to the geomorphological attempts to statistically relate observed tree networks to the predictions of possible models of their evolution, very little attention has been paid to the statistical analysis of circuits, exceptions being the work of the statistician Ling (1973), summarized in Getis and Boots (1978, p. 104) and Tinkler (1977).

As before, we can develop a random topological model as the appropriate starting-point and compare its predictions with any observed network with the same number of nodes and links. Since a circuit network can be regarded as a tree network with a major constraint removed, it comes as no surprise to note that the number of structurally distinct circuits that can be created from n nodes rises very rapidly indeed as n is increased. In fact, to make progress, it is necessary to specify not only n, the number of nodes, but also q, the number of links to be assigned between them. Our test statistic will be based upon the frequency distribution of the number of paths incident at the nodes, or *nodality*, given by the sum of each row of a 0/1 connectivity matrix.

Intuitively, it is possible to envisage a number of possible frequency distributions of nodality that depend upon the type of network involved. For example, a network of paths between central places of equal rank will give a highly connected graph in which places tend to have similar nodality, giving a very peaked frequency distribution of low variance. In contrast, a network representing shopping trips might show a concentration of paths on a single node, giving a frequency distribution with high variance and a wide spread of nodalities. What would the long-run expectation of nodalities be, if we postulated a random allocation of paths between nodes?

In fact, this problem is very similar to the basic binomial model introduced in Chapter 3 to describe the random allocation of points to quadrats. Here we assign links to nodes, but with an important difference. Although the assignment of each 0/1 link between nodes can be done randomly, so at each step all nodes have equal probability of being linked, each placement successively reduces the number of available possible links, hence the probability of a specific link will change. The process is still random, but it now involves *dependence* between placements. If there are n nodes with q paths distributed among them, it can be shown (see Tinkler, 1977, p. 32, in conjunction with Lewis, 1977, p. 67) that the appropriate probability distribution taking this dependence into account is the *hypergeometric*:

$$\text{probability } (m;\, n,\, q) = \binom{q}{m}\binom{N-q}{n-1-m}\bigg/\binom{N}{n-1}$$

in which n and q are as above, m is the required nodality value, and N is the total number of possible links in an n-node circuit network. In a symmetric connectivity matrix that describes the graph, N is the number of distinct entries, that is

$$N = n(n-1)/2$$

We can illustrate the use of this probability distribution using the Skye network data originally presented in Figure 4.7. There are $n = 9$ nodes with $q = 10$ paths, so that the total possible number of links is

$$N = 9(9-1)/2 = 36$$

The probability distribution of nodality values from $m = 0$ (unconnected node) through to $m = n - 1$ (a node directly connected to every other) beginning with $m = 0$ is found as follows:

$$\text{probability } (0;\, 9,\, 10) = \binom{10}{0}\binom{36-10}{9-1-0}\bigg/\binom{36}{8}$$
$$= \frac{10!\ 26!\ 8!\ 28!}{0!\ 10!\ 8!\ 18!\ 36!} = 0.0516$$

(To evaluate these factorials, remember first to cancel out wherever possible, then resort to Stirling's approximation.) Similarly, the probability of nodality values of $m = 1$ is:

$$\text{probability } (1;\, 9,\, 10) = \binom{10}{1}\binom{36-10}{9-1-1}\bigg/\binom{36}{8} = 0.2174$$

and so on. The complete set of values is listed in Table 4.10, which also presents the results of a significance test against the observed nodalities. Using the Kolmogorov-Smirnov test (see pp. 56–8), it can be seen that the maximum value of the differences in cumulative proportions at 0.064 is much less than the critical value at the 95 per cent confidence level found from tables to be 0.624. We do not have enough evidence to be able to reject the null hypothesis of random linking. The reader should, of course, note that this does not imply that the islanders of Skye built their roads randomly!

Table 4.10 Kolmogorov-Smirnov significance test of the frequency distribution of nodality values for the Skye data against a hypergeometric alternative

Degree of nodality m	*Hypergeometric probability*	*Observed frequency*	*Observed cumulative*	*Expected cumulative*	*Binomial probability*
0	0.0516	0	0.0000	0.0516	0.0810
1	0.2174	3	0.3333	0.2690*	0.2315
2	0.3424	2	0.5555	0.6114	0.2976
3	0.2608	3	0.8889	0.8722	0.2267
≥4	0.1278	1	1.0000	1.0000	0.1632
	1.0000	9			1.0000

$d_{\max} = 0.0643$
$d_{\text{crit}} = 0.6240$

* Maximum number.

A few minutes' work with a hand calculator will convince you that the hypergeometric distribution presented above is not very easy to handle when n (and hence N) gets large. Fortunately, as n increases, so it becomes possible to use the simpler binomial distribution originally introduced on p. 52 and which, it will be recalled, assumes an independent random allocation. This is understandable, because if n is large, the hypergeometric probabilities will not change much as the nodes are connected, so that it becomes reasonable to simplify by assuming that they do not change at all. Using the notation we have used above, this binomial can be written:

$$\text{probability } (m;\, n,\, q) = \binom{q}{m}\left(\frac{2}{n}\right)^{m}\left(1 - \frac{2}{n}\right)^{q-m}$$

If you remember that at each allocation of a path *two* nodes are joined, then the probability of a path terminating at a node is:

$$p = \frac{2}{n}$$

and that the probability of there not being a link is:

$$q = 1 - \frac{2}{n}$$

This expression is exactly the same as that developed on p. 52. As the final column of Table 4.10 shows, even with n as low as 9, the approximation to the true hypergeometric probabilities is a reasonable one.

SOME CONCLUSIONS

This chapter has covered a great deal of ground and, seemingly, has introduced what might be seen to be some quite difficult mathematics. We have introduced the ideas of a matrix, vector resolution using trigonometry, matrix multiplication, continuous and discrete probability distributions and, finally, some complex expressions to establish numbers of TDCNs and circuit networks. There are two general conclusions that the reader should carry forward. First, it should have become obvious that although it is easy to declare the existence of fundamental spatial concepts like pattern, distance, direction and connection revealed when we map line data, these are surprisingly difficult to capture analytically. Even the simplest of these notions involves advanced mathematical concepts. Second, although the precise details have varied, the basic procedure for statistical testing our line data has been the same as it was for points in Chapter 3. We have postulated a process – in all cases a random process – and predicted the probabilities of particular, specified realizations of that process. Statistical theory as such has entered only when we have attempted to compare our observed patterns with this randomly generated standard. Notice that even if we have to reject the random process in favour of one that produces patterns either more or less structured than random, we are still some way from specifying this alternative.

RECOMMENDED READING

General texts

Getis, A. and Boots, B. (1978) *Models of Spatial Processes*, Cambridge, Cambridge University Press, especially Chapter 5, 86–120.

Haggett, P. and Chorley, R. J. (1969) *Network Analysis in Geography*, London, Arnold.

Lewis, P. (1977) *Maps and Statistics*, London, Methuen.

Siegel, S. (1956) *Non-Parametric Statistics for the Behavioral Sciences*, New York, McGraw-Hill. (This provides details of the Kolmogorov-Smirnov test used extensively in this chapter.)

Taylor, P. J. (1977) *Quantitative Methods in Geography*, Boston, Houghton Mifflin.

Wilson, A. G. and Kirkby, M. J. (1980) *Mathematics for Geographers and Planners*, 2nd edn, Oxford, Oxford University Press.

Simple line data

Baxter, R. S. (1976) *Computer and Statistical Techniques for Planners*, London, Methuen.

Curray, J. R. (1956) 'The analysis of two-dimensional orientation data', *Journal of Geology* 64, 117–31.

Dacey, M. F. (1967) 'Description of line patterns', Evanston, Ill., Northwestern University Studies in Geography 13, 277–87.

Durand, D. and Greenwood, J. A. (1958) 'Modifications of the Rayleigh test for uniformity in analysis of two-dimensional orientation data', *Journal of Geology* 66, 229–38.

Horowitz, M. (1965) 'Probability of random paths across elementary geometric shapes', *Journal of Applied Probability* 2, 169–77.

Mardia, K. V. (1972) *Statistics of Directional Data*, London, Academic Press.

Nordbeck, S. (1964) 'Computing distances in road nets', *Papers, Regional Science Association* 12, 207–20.

Pincus, H. J. (1956) 'Some vector and arithmetic operations on two-dimensional orientation variates with application to geological data', *Journal of Geology* 64, 553–7.

Rayner, J. N. (1971) *An Introduction to Spectral Analysis*, London, Pion.

Speight, J. G. (1965) 'Meander spectra of the Angabunga river', *Journal of Hydrology* 3, 1–15.

Taylor, P. J. (1971) 'Distances within shapes: an introduction to a family of finite frequency distributions', *Geografiska Annaler* B53, 40–54.

Timbers, J. A. (1967) 'Route factors in road networks', *Traffic Engineering and Control* 9, 392–4, 401.

Tree networks

Gardiner, V. (1975) *Drainage Basin Morphometry*, British Geomorphological Research Group, Technical Bulletin 14, Norwich, Geo Abstracts.

Gardiner, V. and Park C. C. (1978) 'Drainage basin morphometry: review and assessment', *Progress in Physical Geography* 2, 1–35.
Haggett, P. (1967) 'On the extension of the Horton combinatorial algorithm to regional highway networks', *Journal of Regional Science* 7, 282–90.
Horton, R. E. (1945) 'Erosional development of streams and their drainage basins: hydrophysical approach to quantitative morphology', *Bulletin, Geological Society of America* 56, 275–370.
Jones, J. A. A. (1978) 'The spacing of streams in a random walk model', *Area* 10, 190–7.
Maxwell, J. C. (1955) 'The bifurcation ratio in Horton's law of stream numbers', *Transactions, American Geophysical Union* 36, 520.
Milton, L. E. (1966) 'The geomorphic irrelevance of some drainage net laws', *Australian Geographical Studies* 4, 89–95.
Riddell, J. B. (1973) 'An expansion of the Horton stream ordering model to circuited transportation networks', *Geographical Analysis* 5, 351–7.
Shreve, R. L. (1966) 'Statistical law of stream numbers', *Journal of Geology* 74, 17–37.
Strahler, A. N. (1952) 'Dynamic basis of geomorphology', *Bulletin, Geological Society of America* 63, 923–38.
Werritty, A. (1972) 'The topology of stream networks', in Chorley, R. J. (ed.) *Spatial Analysis in Geomorphology*, London, Methuen, 167–96.

Circuit networks

Carter, F. W. (1969) 'An analysis of the medieval Serbian oecumene: a theoretical approach', *Geografiska Annaler* B51, 39–56.
Garner, B. J. and Street, W. A. (1978) 'The solution matrix: alternative interpretations', *Geographical Analysis* 10, 185–90.
Garrison, W. L. (1960) 'Connectivity of the interstate highway system', *Papers, Regional Science Association* 6, 121–37.
James, G. A. *et al.* (1970) 'Some discrete distributions with applications to regional transport networks', *Geografiska Annaler* B52, 14–21.
Kansky, K. J. (1963) 'Structure of transport networks: relationships between network geometry and regional characteristics', Chicago, University of Chicago, Department of Geography Research Papers 84.
Ling, R. F. (1973) 'The expected number of components in random linear graphs', *Annals of Probability* 1, 876–81.
Ord, J. K. (1967) 'On a system of discrete distributions', *Biometrika* 54, 649–56.
Pitts, F. R. (1965) 'A graph theoretic approach to historical geography', *Professional Geographer* 17, 15–20.
Tinkler, K. J. (1977) 'An introduction to graph theoretical methods in geography', Norwich, Geo Abstracts, CATMOG 14.

WORKSHEET

(1) Examine the papers by Garrison (1960), Kansky (1963), Pitts (1965) and Carter (1969) and write an account of their use of network analysis to solve substantive research problems.

(2) List all the line symbols used on an Ordnance Survey 1:50,000 topographic map or a US Geological Survey 1:62,500 sheet and comment upon their effectiveness.

(3) On pp. 70–1 a rule was introduced that to be legible a line symbol must exceed an angle of two minutes of arc at the eye. In collaboration with others, design an experiment using different line widths, map readers and viewing distances to verify this rule. If it does not seem to work, write up your results and submit them for publication in a cartographic journal.

(4) Obtain a list of road distances between towns, as for example published in the UK Automobile Association's *Members' Handbook*. Measure the straight-line distances between these points and compute the average route factor. Repeat the exercise, using a map of British Rail's inter-city services and compare the result. A similar exercise can be constructed for distances and travel times in North America using the mileage chart and driving time map produced in the Rand McNally *Atlas*.

(5) Using a 1 : 25,000 or 1:24,000 topographic map, find an example of a fourth-order (Strahler) drainage basin. Trace off the drainage network and order it using Strahler's method. Use your results to test the law of path numbers.

(6) A drainage network with four sources has five possible TDCNs. Draw them.

(7) Use the Shreve equations introduced on pp. 99–102 to verify that a drainage network with nine sources has 1430 possible TDCNs with possible Strahler stream-order profiles:

Continued on next page

profile	*number*
N (9, 4, 2, 1)	14
N (9, 4, 1)	56
N (9, 3, 1)	560
N (9, 2, 1)	672
N (9, 1)	128

Use the results to calculate the probability of each order profile and, hence, the most probable average bifurcation ratio. Finally, find a natural nine-source basin and see how nature behaves!

(8) On pp. 87–9 the **C** matrix describing the main road network of Skye with 1s down its principal diagonal was raised to the fourth power. *Either* repeat this analysis using 0s on the diagonal and contrast your results with those given in the text, *or* use the Garner-Street weighting method to reconstruct the same.

CHAPTER FIVE

·AREAS ON MAPS·

INTRODUCTION

Chapters 3 and 4 examined the use of point and line symbols to represent data having zero and one spatial dimension, respectively. The present chapter extends the analysis to the representation of area data having the spatial dimensions of a length × length L^2, that is with two spatial dimensions. Such data can arise in a number of ways, depending on the level of measurement and the nature of the areas to which they refer, but they introduce several new spatial concepts – notably *area, shape* and *contiguity* – that we might wish to capture analytically. As will be seen, the measurement of *area* is not without problems, *shape* has proved one of the most difficult of all geographical concepts, and it is only recently that progress has been made in the analysis of *contiguity,* the pattern of connections between areas.

Areas of the earth's surface have particular relevance in the development of geographical theory. Most of our knowledge of the world is based upon census information collected over discrete areal units, such as counties and countries, and the recognition and delimitation of regions has been a major practical and academic activity. It is not surprising, therefore, that numerous ways have been developed to map and analyse areal data.

THEORY AND PRACTICE IN AREA SYMBOL MAPPING

As shown in Table 5.1 and Figure 5.1, many different types of maps can be drawn to represent area data. From the point of view of the data themselves, the by now familiar distinction between nominal, ordinal, and interval and ratio scaled data, can be maintained. The attribute of area is more complex, but a useful way of subdividing is to ask how the areas were defined. On the one

Table 5.1 Types of area-based maps

Measurement level	*Nature of area*	
	Natural	*Imposed*
nominal and ordinal	chorochromatic	colour maps
interval and ratio	not often used 'natural choropleth'	choropleth maps

hand, are true natural areas such as that of woodland on a land-use map or the surface outcrop of a particular geological formation. These can be contrasted, on the other hand, with artificial areal units imposed by the investigator such as census tracts, grid squares, and so on.

The simplest map is the colour patch or *chorochromatic* map, in which some symbolism is used to indicate the presence of a named attribute over a natural or imposed area, as in the three examples in the top left of Figure 5.1. One might, for example, shade in all the areas of the UK classified as being urban in character, or all rock outcrops of Silurian age, or even rainfall in excess of a specified threshold. Very often, chorochromatic maps use several colours to show a number of nominal classes simultaneously (e.g. rocks of differing character), but the simplest of all would portray one nominal category in one colour, giving a *two-phase mosaic* or *binary* map. Sometimes this kind of map is referred to as a *two-colour* map, because one colour is used to indicate the presence of an attribute and, by implication, the unshaded or white areas are a second colour, indicating where the attribute is not found. In general, we can talk of a '*k*-colour' map, where *k* is the number of colours involved. Chorochromatic maps using natural areas give a very clear picture of distribution and are unlikely to be misinterpreted, but the same cannot be said of those based on data collected over a set of imposed areas. Without further information, all that this second type of map tells us is that the attribute is present somewhere in the area, or, in aggregate, the area is of the type implied. An example would be that shown in the top right of Figure 5.1, indicating the political colour of a series of parliamentary constituencies. The resulting area affiliations can depend as much on the artificial boundaries used as on the underlying distribution of voting behaviour.

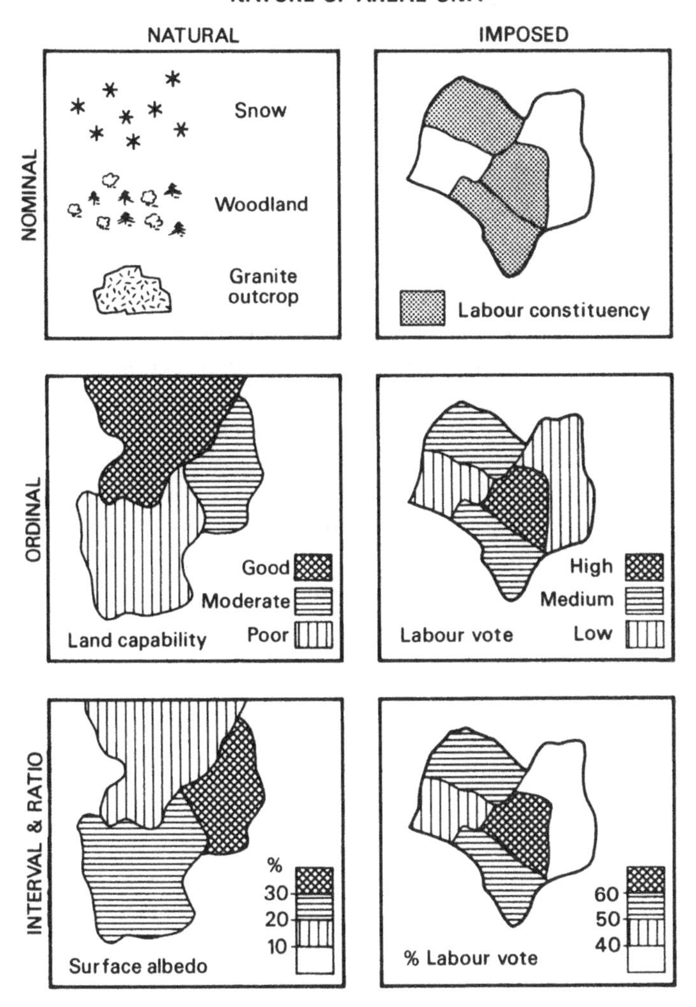

Figure 5.1 Some examples of mapping area data.

A second problem occurs if the nominal categories used, although mutually exclusive in themselves, are not spatially exclusive. On a map the distributions will overlap. Faced with this problem, several solutions can be adopted. An obvious one is to choose and map only those categories which are spatially exclusive; others are to interdigitate sets of point symbols, to use special symbols or colours for the mixed distributions, or simply to draw the outline of each.

The symbolism used on colour maps can be quite complex, with a choice of both symbols and colours. The simplest approach is to use a wash or shading over the area concerned, but an attractive alternative is to overlay a repetition of simple, meaningful point symbols, for example, the snow and woodland in Figure 5.1. Using a colour wash ensures that accuracy of the outline is maintained, but if symbols are used, the accuracy of the edges will be limited by the symbol size.

As the name suggests, chorochromatic maps are often drawn using colour, but colour selection – even for relatively simple representations – is one of the most complex of all cartographic problems, involving considerations of reproduction methods and costs, established convention and the user's reaction. Used successfully, colour can give enormous gains in clarity and legibility, and even before the days of colour printing, maps were often hand coloured. Equally, if used badly, colour can confuse and is still expensive to reproduce; for research publication it is usually better avoided. From the point of view of human reactions, colour has three dimensions – hue, value, chroma. *Hue* refers to that sensation which we describe as red, blue, green, and so on: its specification is rather complex, but a widely available international standard is the well-known Munsell notation, developed in 1915. What is evident is that our sensitivity to colour varies according to hue, being most of all for red, followed by green, yellow, blue and purple. Extreme care, then, should be taken if map colours are intended to represent a graded series. Second, convention plays a major part in the choice of colour. We associate blue with water, green with woodland, red with warmth, blue with cold, yellow with desert, and so on. *Value* is the sensation of lightness or darkness given, the important rule being that the higher the value being represented, the darker the value; but notice that different hues also have associated with them different intrinsic values, red always seeming darker than yellow, for instance. Finally, *chroma* refers to the amount of a hue in a colour, that is its apparent intensity or brilliance. Again, it is important to ensure that the greater the intensity, the greater the magnitude being mapped; but a complication is that chroma effects vary according to the size of the coloured areas. Although a small area coloured a brilliant red can be acceptable, the same colour used over a large area could well prove totally overpowering.

A second type of area-based map is one which represents ordinally scaled data, as illustrated in the middle row of Figure 5.1: for example, 'good', 'moderate' and 'poor' natural areas on a scale of land capability; or 'high', 'medium' and 'low' proportions of the vote for a political party over a series of imposed electoral districts. Representing such a series is much the same as outlined already, except that great care must be taken to ensure that the colours or symbols used correspond to the ordinal gradations. Several techniques can be used. Appropriate point symbols drawn at differing densities are one possibility, another is to use differing shading styles (as in Figure 5.1) and, if colour is to be used, the value of a single hue may be varied from light to dark. At greater risk of cartographic disaster, different hues could be employed.

The most complex of all maps of area-valued data are those for interval or ratio scaled data collected over imposed, or less commonly, discrete natural areas. A typical example of an imposed area map is that in the lower right of Figure 5.1, which shows the distribution of voting behaviour measured as a percentage of the total vote over a series of parliamentary constituencies. These areas are clearly artificial, imposed ones, but it is possible to imagine similar representations of natural area-valued data, such as the total rainfall in each of a series of drainage basins or, as the lower left of Figure 5.1 shows, the percentage surface albedo in differing vegetation areas. A common way of representing both types of area-valued data is by a *choropleth* map, in which a graded series of shades is used to indicate increasing values of the variable. Of necessity, the shading is applied over all the area to which it applies and a key to the range of values assigned to each shade must be presented.

The procedure, illustrated in Figure 5.2, is deceptively simple. Initially, the data for each area are collected as in (a) in the figure, which shows the percentage of the total 1971 population of each ward of the city of Leicester coming from the New Commonwealth. Second, a graded series of shades is devised and each areal unit assigned to its appropriate category as in Figure 5.2(b). Finally, and unless there are good reasons for not doing so, it is usual to delete all the superfluous area boundaries over which the shaded category is the same, as in Figure 5.2(c). Because geographical data, especially from census sources, are agglomerated to some sort of areal unit, choropleth maps are common in virtually all human geography research. The pattern revealed in Figure 5.2(c), for example, poses a great many questions about the social processes that generated it.

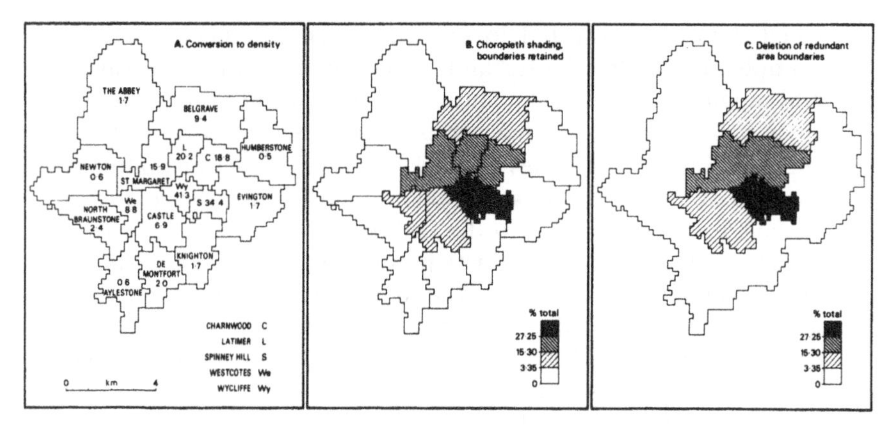

Figure 5.2 a, b and c Drawing a choropleth map.

This apparent simplicity of the choropleth map should not blind us to its many obvious and not so obvious deficiencies including the area-dependence of the data, the perceptual and statistical effects of using different sizes of areal unit, the danger of making inferences about aggregates of people that are not warranted at an individual level, and the effects of choice of class number and interval. As will be seen, these problems are usually intrinsic facts of life over which the researcher has no control, but it pays to be aware of them.

Area dependence

At the outset, it has to be realized that any apparent pattern, such as that in Figure 5.2(c), revealed when area-valued data are mapped may be as much a result of the zones chosen as of the underlying distribution. It is possible to design a system of zones that would completely remove the apparent high densities of recent immigrants in the Wycliffe and Spinney Hill wards of Leicester. It is equally possible to design zones that would lead to even higher densities being recorded, the critical factor being the relationship between the real house by house pattern on the ground and the boundaries of the zones imposed.

This effect may be illustrated by a simple experiment. Figure 5.3(a) shows a set of thirty-six basic spatial units (e.g. houses, factories, or whatever) to which have been allocated randomly selected numbers in the range 0–9. Next, in Figure 5.3(b), we

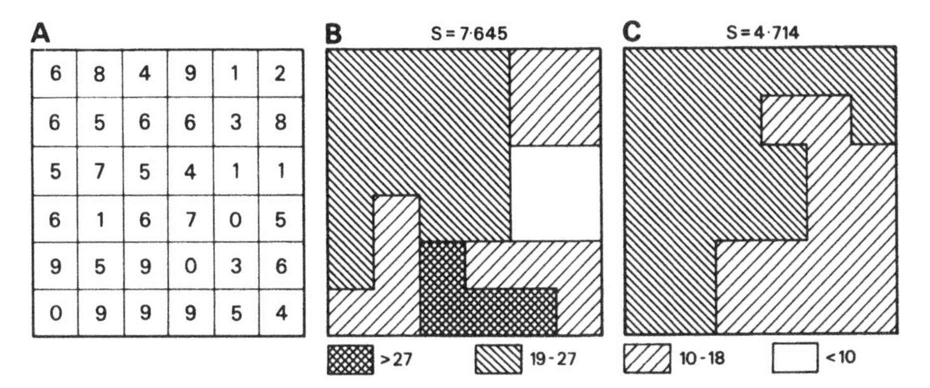

Figure 5.3 a, b and c Illustrating the area-dependence problem.

overlay a series of nine zones each of area equal to four of the basic units, count the total in each zone, and assign it to one or other levels of a graded series of choropleth shades. There is an apparent concentration of high values on the southern edge of the mapped area. Figure 5.3(c) shows a different set of equal-area zones which completely removes the concentration, and which reduces the standard deviation of the mapped values from 7.65 to 4.71. In this example, equal-sized areas are used, but if the areas are also varied, these effects can be even more marked. Geographical data are almost invariably collected over irregular and unequally sized zones, frequently bearing little relationship to underlying patterns. In a careful analysis, Coppock (1955, 1960) has, for example, shown how inappropriate is the civil parish for reporting UK Agricultural Census data which are collected at the farm level. Individual farms can stretch over more than one parish, parish boundaries themselves frequently take in very varied and dissimilar agricultural areas and show an enormous range in size. The effect of zone area on any statistical analysis conducted using zone data has been studied by Robinson (1956). His simplified example is shown in Figure 5.4, on which three alternative zoning schemes have been laid out. In (a) in the figure there are six zones and the bracketed values are the densities of two variables x and y. The product-moment correlation coefficient is 0.75. In (b) in the figure, one of the areas has been doubled and the density recalculated, but the resulting correlation coefficient is now 0.875, and in (c) with an even larger area of uniform density, there is a correlation of 0.500. It is clear that whatever the underlying true correlation, the

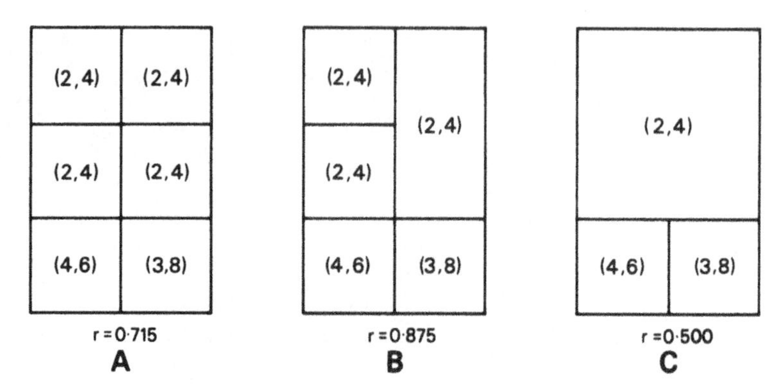

Figure 5.4 a, b and c Illustrating the size-dependence problem.

imposition of data-collection zones leads to very different values for the correlation coefficient. It is, in fact, possible to get virtually any value we like for any statistic, simply by juggling with the zone boundaries, a fact long known to politicians because it forms the basis of the gerrymander, drawing electoral boundaries to suit one's own political affiliation.

A second problem of area-based data occurs when we attempt to relate two or more distributions, either visually or statistically. Suppose, for example, that in a study of recent immigration into Leicester (Figure 5.2) we observe a high correspondence between a percentage map of immigrants from the New Commonwealth and a map of crime rates. The temptation is to transfer this macrolevel correspondence to the individual microlevel and assert that immigrants are more likely to be criminals. In fact, the reverse is almost certainly the case. All areal correspondence shows is that immigrants live in areas where the crime rate is high, and our original assertion is an example of an *ecological fallacy,* that is making an assumption that a relationship which is apparent at a macrolevel of aggregation must also exist at the microlevel.

It should be evident that choropleth maps are dangerous tools that can lead to very deceptive conclusions, but are there any solutions to these area-dependence problems? One attractive solution is to standardize data-collection areas to a regular, equal-sized grid, such as the 1 km^2 of the Ordnance Survey. This ensures that area size is not a problem, but it does nothing to correct for effects due to the location of grid boundaries or the danger of ecological fallacies. More complex methods accept the zone areas as a fact of life and modify the statistical analysis accordingly. One simple procedure, proposed by Robinson (1956), is to weight density values by the zone area to which they refer prior to calculating any other statistics (see Robinson, 1956; Robinson *et al.*, 1961, for details). A second solution is attempting to even out irregularities in the collecting zones by careful and sensible amalgamation of adjoining units, as illustrated by Coppock (1955, 1960) in his use of agricultural data.

The area-dependence problem of geographic data is therefore severe, with many possible sources of error in assembling and interpreting choropleth maps, but recent research has begun to develop an approach which may go some of the way to reducing these errors. Instead of thinking of any given pattern of zones as fixed for all time, this considers the act of zone-definition as a type of sampling procedure.

It is often asserted that geographic census data, aggregated over imposed and fixed zones, are not appropriate for statistical analysis simply because they are in no sense a sample from an underlying population of values. The argument is that they *are* that complete population, so while it is legitimate to calculate descriptive statistics, it is pointless estimating their statistical confidence limits or in testing significance; the values calculated are the true parameter values. As recent work has shown, this argument only holds if we think of zones as fixed. As we have seen, changing zones leads to changes in the values of the statistics obtained, so that it is permissible to think of a distribution of values obtained under differing zone designs. It is possible, therefore, to regard zoning as a form of sampling and refer to the *zoning distribution* of a particular statistic in much the same way as a statistician refers to a sampling distribution. As yet, little progress has been made in the specification of zoning distributions.

Class number and interval

We have just demonstrated that the appearance of a pattern in our map of immigrants in Leicester (Figure 5.2(c)) depends to a large extent on the zone boundaries used to aggregate the data. A second source of dependence is in the choice of the number and class limits for the shade categories used. The map (Figure 5.2(a)) on which we wrote the actual numbers is honest in that each zone has its correct value, but Figure 5.2(c) generalizes these data by imposing four shades, each spanning an equal range of values. The selection of these shades is an important factor in determining the appearance of the final map, and an enormous literature, summarized by Evans (1977), has built up giving guidelines on how to proceed.

An initial decision to be made is why have class intervals at all? Modern technology can produce graded series of shades with as many, or very nearly as many, grades as there are data values. One way to do this was suggested by Tobler (1973) and involves continuously varying the distances between the lines used to shade each map zone. If a value is high, the lines are drawn close together; if it is low, they are drawn wide apart. In principle, this type of 'choropleth map without class intervals' could be hand-drawn, but in practice, a computer and computer-plotter would be used. A second method uses facsimile reproduction equipment, which usually allows up to 255 grey tones on a continuum from black to white (Muller and Honsaker, 1978). At first sight class-interval-free choropleth maps seem an attractive answer to a difficult problem, but as Dobson (1973) points out, they may be no solution at all. First, no matter how technically capable we are, it can be argued that it makes little sense to use more map shades than the eye can easily distinguish, and it is generally held that this ranges from 4 to 10 with a good average of 7 or 8. Second, many cartographers insist that each zone should be unambiguously seen in its correct value class, implying that the shades used must be visually separable. Finally, interval-less choropleths do not generalize – a 'choropleth map without class intervals' does not need a computer for its production. All we need to do is write the value in each zone! On balance, then, it seems better to impose a series of shades on the data; the practical problem is to decide how many, spanning what intervals and with what type of shading.

Number of categories. The choice of number of classes to use depends on the nature of the intended audience, the available technology, the number of zones to be mapped and the frequency distribution of the data. It is generally agreed that with ten classes, most users would have difficulty in reading the map, yet, at the opposite end of the scale, too few classes can overgeneralize. An unsophisticated audience, poor mapping technique, a small number of zones, and a highly bimodal frequency distribution of data, all argue for few class intervals. A sophisticated audience, good mapping tools, a

Table 5.2 A classification of class-interval systems

(1) *Exogenous:* meaningful in relationship to threshold values which are not derived from the mapped data, such as a sex ratio of 1.

(2) *Arbitrary:* numbers of no particular significance often with an unequal class interval, such as 5, 10, 20, 30, 80, 120. . . .

(3) Various *ideographic* systems affected by the details of the data to be mapped, such as:

- (a) *multimodal,* using 'natural breaks' in the data frequency distribution;
- (b) *multistep,* with breaks where the slope of a cumulative frequency curve changes;
- (c) *contiguity-based,* so as to maximize the extent and minimize the number of regions with a given class shade;
- (d) *correlation-based,* so as to maximize similarity to a given map;
- (e) *percentile classes,* which contain equal numbers of spatial zones or near-equal zone areas;
- (f) *nested means class limits,* where the frequency distribution mean is taken as threshold for subdivision into two classes, then each class is subdivided about its mean to give four classes, and so on;

(4) Various *serial* schemes, with limits in a definite mathematical relationship to each other, such as:

- (a) *normal percentiles,* with class limits related to equal-frequency classes on an assumed normal curve;
- (b) *standard deviation units,* centred on the mean which is a class midpoint if the number of classes is odd, and a class boundary if it is even;
- (c) *equal intervals;*
- (d) *equal intervals* on a reciprocal scale;
- (e) *equal intervals* on a trigonometric series;
- (f) *geometric progression* of class widths;
- (g) *arithmetic progressions;*
- (h) *curvilinear progressions,* where a plot of the log of class limit against log of class number is a smooth curve.

large number of zones, and a uniform frequency distribution of data, argue for a large number of zones, at least six or seven. Figure 5.5(a)–(d) shows the distribution of public-owned housing by wards of the city of Leicester in 1971, drawn using two, four and eight equal-interval classes. All show that this housing is found in zones away from the city centre, but it is arguable which shows this most clearly.

The class-interval system. The second problem concerns the class-interval system to be used, and a very large number of possible schemes have been suggested. Table 5.2 lays out the sixteen schemes recognized by Evans (1977). It is clear from this list that

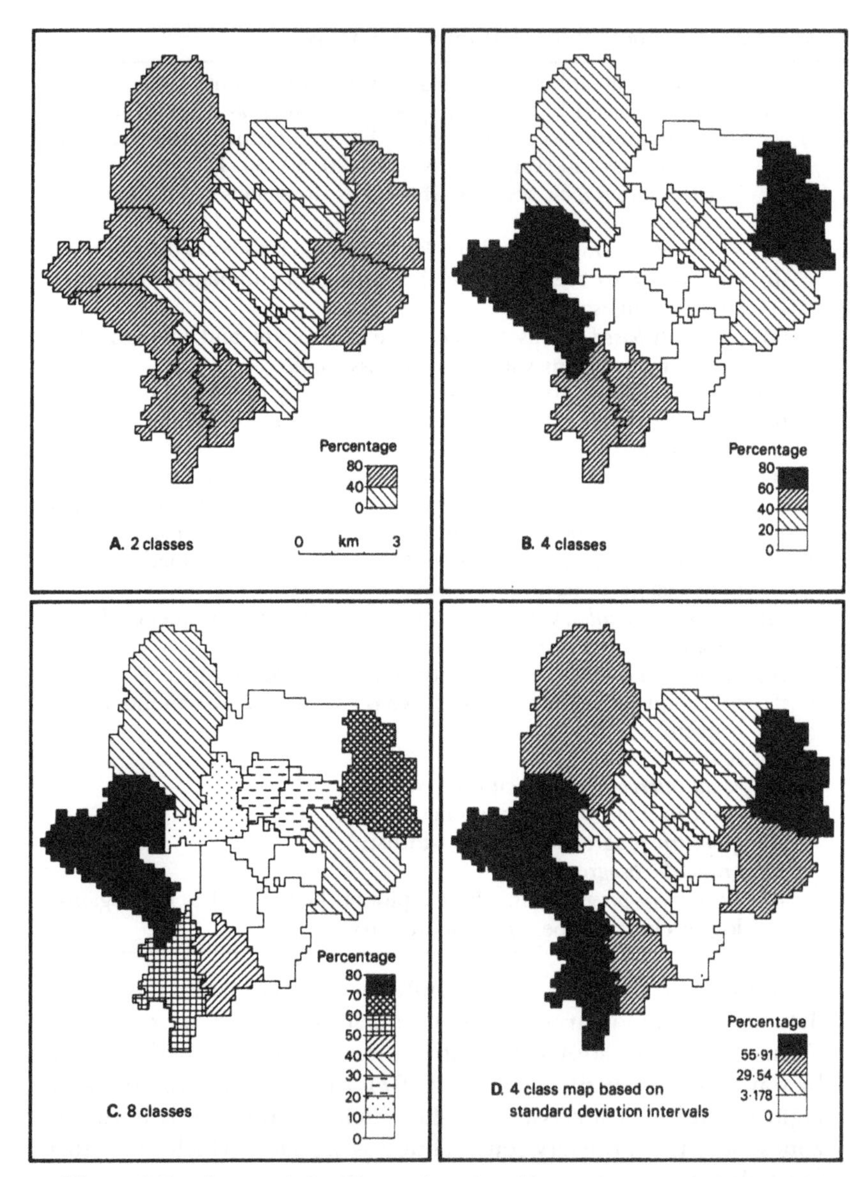

Figure 5.5 a, b, c and d Illustrating the effects of number of classes.

the choice of scheme will again very much depend upon the sophistication of the intended user and the frequency distributions of the data. As a general rule, Evans recommends intervals based on standard deviation units if the data are normally distributed, on equal and regular intervals for uniform distributions, and geometric progressions for J-shaped distributions. Data that are multimodal should have class limits arranged to take account of natural breaks in the frequency distribution.

How to shade. It is not enough to decide simply upon the number of classes and class-interval systems to be used. There must also be a careful choice of the actual shades used and the legend to accompany the completed map. It is important to find a gradual series in which there is sufficient contrast between steps to make each distinguishable. As a general guide, contrasts in value among patterns can be obtained by varying the percentage of the area inked either by a network of lines, or a stipple of dots. Lines are easy to apply but often obscure lettering, and may produce irritating side-effects, whereas dots are visually clearer but much harder to apply (see Williams, 1958; Jenks and Knos, 1961). Nowadays, it is usual to use preprinted area symbolism, the zone shapes being cut out and stuck on to the final map. In using this or any other shade symbols, remember that photoreduction of the final product will radically alter the appearance, generally by reducing the contrast between shades.

The concept of error on choropleth maps

Figure 5.5(d) shows a four-class map of public-owned housing in Leicester constructed using standard deviation intervals, which should be compared with the equal-interval map shown in (b). The number of classes and symbolism are the same, but which is the better map? Ultimately, this is a question whose answer depends upon the use to which it is put, involving very complex considerations of the user's perception, but some progress towards a mathematical definition of 'better' has been made by Jenks and Caspall (1971) in their concept of *error* on a choropleth. This is based on the simple idea that when the data are grouped into classes, this obscures the individual zone values, replacing each by one of a limited number of shades. The shades are a generalization of the underlying true values and the differences between the generalized value and the true value summed over all zones is a measure of the map error. The most accurate map we could draw is a choropleth without class intervals in which shades exactly represent the data. The least accurate would be one with just one category drawn through the mean of the data.

Jenks and Caspall proposed three indices of error but the index described below, although it is based on the same idea, differs from their indices in some important respects and is analogous to the percentage reduction in sums of squares used in regression analysis. A specimen calculation for the even-interval, four-class map, in Figure 5.5(b) is shown in Table 5.3. For each zone, the error is calculated as the squared difference between the true value and the mid-point of the category to which the zone has been

Table 5.3 Calculation of an error index for a choropleth map (see Figure 5.5(b)

	(1) *Proportion of area*	*(2)* *Value (%)*	*(3)* *Fitted choropleth value*	*(1)* × $[(2) - (3)]^2$	*Maximum* *(1)* × $[(2) - mean]^2$
The Abbey	0.182	36.9	30	8.67	9.85
Belgrave	0.099	6.7	10	1.09	52.05
Humberstone	0.082	61.1	70	6.49	81.54
Charnwood	0.034	22.7	30	1.79	1.58
Latimer	0.027	20.4	30	2.45	2.22
St Margaret	0.049	14.0	10	0.80	12.05
Newton	0.085	77.0	70	4.17	191.88
North Braunstone	0.062	78.5	70	4.47	148.43
Westcotes	0.029	2.1	10	1.82	21.91
Aylestone	0.062	58.8	50	6.93	52.65
Wycliffe	0.019	3.2	10	0.92	13.85
Spinney Hill	0.025	3.1	10	1.19	17.44
Castle	0.057	8.1	10	0.02	26.38
De Montfort	0.050	44.3	50	1.62	10.86
Knighton	0.055	0.9	10	4.54	45.01
Evington	0.081	34.9	30	1.96	2.34
	1.000	472.7		48.93	690.03

Mean value = 472.7/16 = 29.54%

$$\text{Error index, } E = 1 - \frac{48.93}{690.03} = 0.929$$

assigned. Using a squared difference, rather than the absolute value of the difference used by Jenks and Caspall, removes any negative signs and gives most weight to the more extreme differences. This value is then weighted by the proportion of the total mapped area lying in the zone and the values for all zones are summed to give a fitted error value. For example, in Figure 5.5 and Table 5.3, The Abbey has a value of 36.9 per cent, which puts it in the second shade category having a midpoint of 30.0. The absolute error is therefore (36.9 −30.0) = 6.9 which is squared to give 47.61. Finally, The Abbey occupies a proportion of 0.182 of the total mapped area, so that the weighted index has the value 8.67. The sum of all such values is 48.93 but, in order to compare this map with any others based on the same data, we need some measure that relates it to the maximum possible error. This total error is calculated in the final column of Table 5.3 as the sum of each zone value less the overall data mean, again squared and area weighted. For The Abbey, this is $(36.9 - 29.54)^2\,0.182 = 9.85$. The final index of error is given by:

$$\text{Error index, } E = 1 - \frac{\text{fitted error}}{\text{total error}}$$

$$= 1 - \frac{48.93}{690.03} = 0.929$$

indicating a close fit. Using the error index as the sole criterion, this equal-interval map is a slightly better model than that based upon standard deviation units (Figure 5.5(c), which has an error index of 0.902, and both are much better than the two-category map (Figure 5.5(a)), which has an index of 0.694.

THE ESTIMATION OF AREA AND SHAPE

No matter how they arise, area data have a number of properties that the geographer wishes to measure, including their two-dimensional area, shape and spatial pattern. When analysing a chorochromatic map, we might wish to estimate the area of any one specified colour (for example, woodland on a land-use map), or the average shape of each parcel of area, or their overall pattern. Even if the areas are imposed zones, it is still necessary to find their areas as a basis for density calculations.

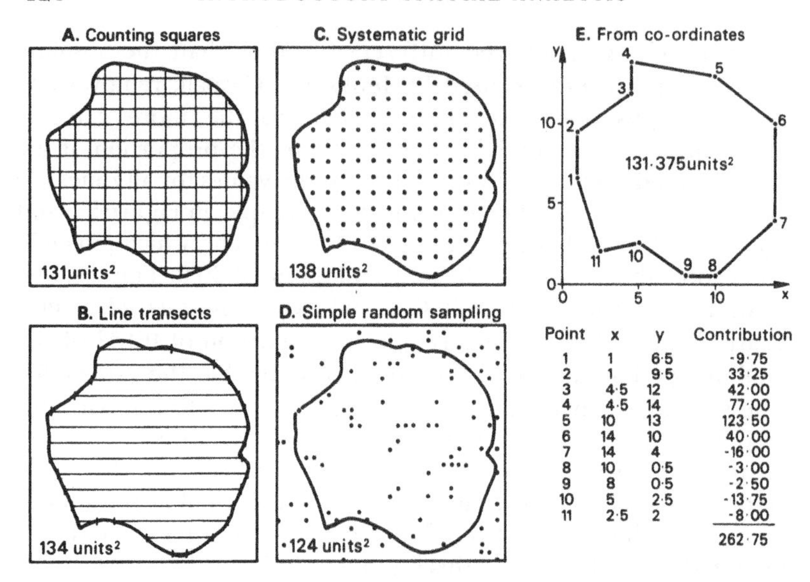

Point	x	y	Contribution
1	1	6·5	-9·75
2	1	9·5	33·25
3	4·5	12	42·00
4	4·5	14	77·00
5	10	13	123·50
6	14	10	40·00
7	14	4	-16·00
8	10	0·5	-3·00
9	8	0·5	-2·50
10	5	2·5	-13·75
11	2·5	2	-8·00
			262·75

Figure 5.6 a, b, c, d and e Methods of measuring area.

Area

Like the measurement of distance examined in Chapter 4, that of area is superficially obvious, but becomes rather more difficult in practice, and a number of methods for area measurement have been proposed ranging from accurate but laborious counting techniques to very rapid but often inaccurate mechanized approaches. The simplest method, illustrated in Figure 5.6(a), makes use of a grid of small squares laid over the map and counting the number of squares which have the greater part of their area in the particular area of interest. It is most efficient to use transparent graph paper for the grid, and the method works well when the measured parcels are few in number and have simple shapes, but large numbers of small areas rapidly give rise to error. A simple alternative, shown Figure in 5.6(b), is to use a network of transects across the area, summing the lengths and multiplying by the line spacing. However, for complex patterns made up of small, irregular parcels, it is generally better to use point counting techniques (see Wood, 1954; Frolov and Maling, 1969). The idea is to replace the square or line with a point and count the number of points which fall into the parcel whose area is required, as illustrated in Figure 5.6(c). A little arithmetic yields the desired estimate of the area. This approach differs radically from square counting in an important respect. It is no longer based on a complete enumeration, but involves the idea of a spatial sample. Usually area measurement and spatial sampling theory are discussed separately, however, they have much in common and many of the sampling designs used in spatial sampling are also appropriate for area measurement (see Berry, 1962; Taylor, 1977). In Figure 5.6(c) a systematic grid sample is used, and in Figure 5.6(d) this is replaced by a simple random sample, in which each point is located using random number tables for both the x and y co-ordinates. Unlike the squares-counting approach, both methods have the advantage that we know quite a lot about the properties of the obtained sample (for a recent discussion, see Switzer, 1976). In a careful comparison, Frolov and Maling (1969) show that both methods break down when small parcels are measured, and recommend a stratification in which the overall point density is varied, increasing as the parcel size decreases. If area measurement is the objective, they further recommend that at least 100 points be counted in each parcel.

A rather different method is shown in Figure 5.6(e). This starts by summarizing the outline of the shape by an irregular polygon with the number of sides chosen as a compromise between the required accuracy and the labour involved. The greater the number of sides, the more accurate the result. Next, the (x, y) co-ordinates of each vertex are recorded, working clockwise from any point, as indicated in the table in the lower half of Figure 5.6(e). The area enclosed can be found from the very simple formula:

$$a = 0.5\sum_{i=1}^{n} y_i \begin{pmatrix} x \text{ co-ordinate} & & x \text{ co-ordinate} \\ \text{of next vertex} & - & \text{of previous vertex} \\ \text{in sequence} & & \text{in sequence} \end{pmatrix}$$

For the first vertex $i = 1$, the previous vertex is x_{11} with a numerical value of 2.5, and the x co-ordinate of the next vertex is 1, giving a contribution to the total summation of

$$6.5\ (1 - 2.5) = -9.75$$

For the second vertex $i = 2$, the contribution is

$$9.5\ (4.5 - 1) = +33.25$$

and so on. The complete sum is 262.75, giving an estimate of the area of 131.375 units^2 which agrees quite well with the other methods. Although this method may seem rather laborious and roundabout, it lends itself very readily to automation and is frequently used in research.

As a few minutes' experimentation will show, all the above methods are exceedingly tedious and it is natural to look for ways of automating the process. One device is the *planimeter*. This has a cursor which is moved around the perimeter of the parcel, giving a direct integration of the enclosed area. Its theory is only moderately complicated (see Gierhart, 1954), but in practice it is extremely difficult to set up, hence a number of other more or less automatic methods have been proposed, including the use of photoelectric cells, cutting out and weighing, using modern table digitizers to record the co-ordinates of the perimeter and, most recently of all, computer-based image analysers, whose theory is much the same as that of counting squares but which do this in a small fraction of the time (Baxter and Lloyd, 1972).

Which method is best? As Figure 5.6 shows and depending on the method adopted, the area of the shape could be anything from 124 to 138 square units. Without further tests, it is impossible to say which is the most accurate. Indeed, it is impossible to give any general rules, largely because the accuracy depends on the spatial characteristics of the areas measured as well as the method used. What is best for one study may prove hopelessly inadequate for another, so that it pays to experiment until a satisfactory method is found.

Shape

Areal units all have two-dimensional shapes, that is constant relationships of position and distance of points on their perimeters. Shape is a fundamental property of many objects of interest in geography, such as drumlins, coral atolls and central business districts. Some shapes, notably the hexagons of central place theory, have important theoretical implications or relationships to postulated generating processes. In the past, shapes were described verbally, using analogies such as 'streamlined' (drumlins), 'ox-bow' and 'shoe-string' (lakes), 'armchair' (cirques), and so on, yet, as Stoddart (1965) showed in a study of the shapes of coral atolls, there was often very little agreement on what terms to use. Although some attempt to quantify the idea of shape seems necessary, in practice most attempts to do this have been less than satisfactory.

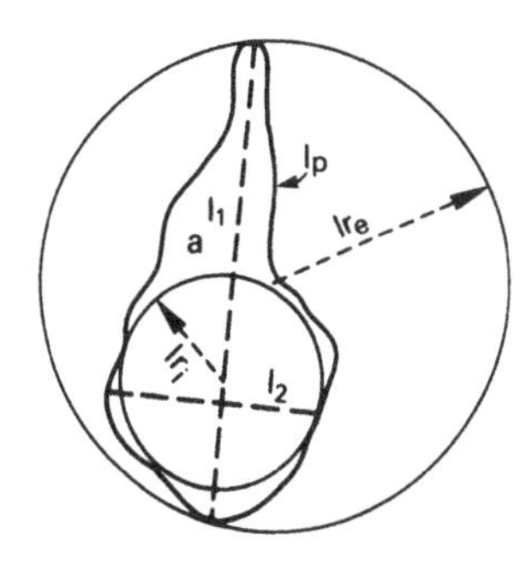

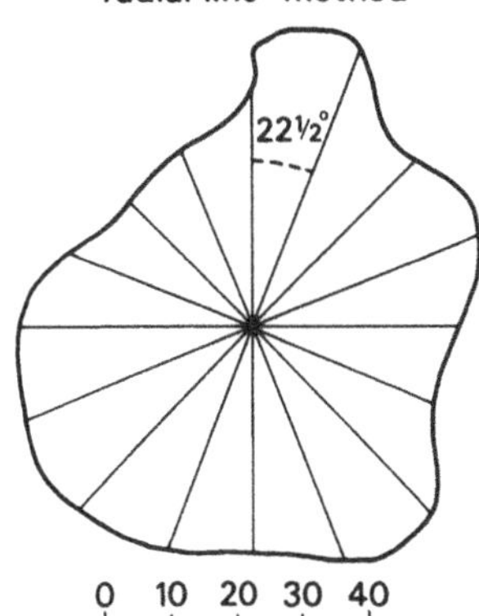

Figure 5.7 a and b Methods of measuring shape.

Given the past use of verbal analogy, the most obvious quantitative approach is to devise indices which relate the given, real-world shape to some regular geometric figure of well-known shape, such as a circle, hexagon or square. Most attempts to date have used the circle. Figure 5.7(a) shows an irregular shape together with a number of possible shape-related measurements that could be taken from it, the perimeter (l_p), area (a), longest axis (l_1), second axis (l_2), the radius of the largest internal circle (l_{ri}) and the radius of the smallest enclosing circle (l_{re}). In principle, we are free to combine these values in any reasonable way but this would not necessarily produce a good index. A good index should have a known value, preferably 1.0, if the shape is circular and, to avoid dependence on the measurement unit adopted, it must also be dimensionless.

One such index is the *compactness ratio*, S_2, defined as:

$$S_2 = (a/a_c)^{0.5}$$

a = measured area of the shape

a_c = area of the circle having the same perimeter l_p as the measured area

which has a value 1.0 if the shape is exactly circular, and has dimension $L^2L^{-2} = L^0$. Other dimensionless ratios which can also be used are the elongation ratio l_1/l_2 and the form ratio a/l_1^2.

A very useful measure devised by Boyce and Clark (1964) is illustrated in Figure 5.7(b) and in Table 5.4. Their *radial line index*

Table 5.4 Calculation of the Boyce and Clark 'radial line' shape index (the distances are from figure 5.7(b), proceeding clockwise from the 0° radial)

i	l_{ri}	$l_{ri}/\Sigma l_{ri}$	*Index values*
1	34	0.061706	0.000794
2	44	0.079855	0.017395
3	35	0.063521	0.001021
4	37	0.067151	0.004651
5	31	0.056261	0.006239
6	30	0.054446	0.008054
7	39	0.070780	0.008280
8	37	0.067151	0.004651
9	33	0.059891	0.002609
10	35	0.063521	0.001021
11	37	0.067151	0.004651
12	37	0.067151	0.004651
13	36	0.065336	0.002836
14	30	0.054446	0.008054
15	27	0.049002	0.013498
16	29	0.052632	0.009868
	551	1.000001	0.098273

$S_2 = 100\ (0.098273) = 9.8273$

operates directly on the shape rather than on one or two measures taken from it. It compares the observed values of distances on a series of n regularly spaced radials originating from a node chosen at the centre of the shape with those that a circle would have, and is defined as:

$$S = 100 \sum_{i=1}^{n} \left| \left(\frac{l_{ri}}{\sum l_{ri}} - \frac{1}{n} \right) \right|$$

The notation | | means the absolute value of the difference is taken, regardless of sign and multiplication by 100 is simply a scaling factor to give the required numerical range. The term within the brackets is the proportion of the total summed length of radii (Σl_{ri}) that each individual l_{ri} represents less the constant value that would be expected for a circle with the same Σl_{ri}. Table 5.4 illustrates the required calculation for Figure 5.7(b). The sixteen radii ($n = 16$)

range in length from 27 to 44 units, and the fixed value for a circle is 1/16 = 0.0625. Hence, the contribution from the 0° radial to the final value is:

$$| (34/551) - 0.0625 | = 0.000794$$

and so on, yielding a final value of 100 × 0.098273 = 9.8273. By definition, a perfectly circular shape would have the value 0, a straight line 200, but in practice values greater than 60 are rare, indicative of considerable elongation.

Although this index has been used in a number of studies, reviewed in Cerny (1975), it suffers from three sources of ambiguity. First, no guidance is given on where to site the central point and most investigators have used the shape's centre of gravity. However, on theoretical grounds it might often be preferable to use some reference point like the town centre or the major city in the country. Second, the choice of number of radii is important. Too few makes the index open to overmuch influence from extreme points on the perimeter, too many and the work of calculation becomes excessive. As the simulation study by Cerny (1975) shows, Boyce and Clark's original choice of sixteen radials at 22.5° intervals seems a reasonable compromise. Third, it is apparent that a great many visually quite different shapes could give the same S value. Even if we confine ourselves to the same sixteen distances used in the example, and remembering that the index value would be the same whichever order we had taken them, we could produce $n!$ different shapes with the same index simply by reordering the radials. In practice, this does not seem to have been too serious a problem. What is important is to ensure that any shape comparisons using this index all involve precisely the same choice of centre and number of radials.

PROPERTIES OF AREA PATTERNS

So far attention has been concentrated solely on the measurable properties of areas as individual units of study without reference either to the overall pattern they create, or to the values that are mapped within them, yet it is frequently these aspects which are of most geographical interest. This section considers both problems, beginning with the study of pure area patterns, and ending with an examination of the important concept of spatial autocorrelation.

Area patterns and contact numbers

Sometimes, as in electoral geography or in geomorphology and biogeography, the patterns made by areas are of interest in their own right, irrespective of any values that might or might not be assigned to them. Such patterns can be as regular as a chessboard, honeycomb, or contraction cracks in basalt lavas, or as irregular as the counties of England and the states of the USA; but to date very few geographers have shown interest in their study. One possible approach is to use the idea of a connectivity matrix **C** (developed on pp. 85–6), but instead of recording the presence or absence of links between nodes, recording contacts between the areas. A contact is said to exist where the areas in question share a common boundary. Entries in **C** can be simple binary 0/1s or some measure of the length of shared boundary. Matrix powering of this *adjacency matrix* proceeds in the same way (pp. 87–93) with results which have modified, but similar interpretations.

A simpler approach which gives less information is to assemble the frequency distribution of *contact numbers,* that is, the number of areas which share a common boundary with each area. An example is given in Table 5.5, which shows the frequency distribution of contacts for the conterminous states of the USA and the counties of

Table 5.5 Contact numbers for the conterminous states states in the USA and the counties of England

Contact number m	*Percentage of conterminous states of USA*	*Percentage of English counties*	*Percentage from independent random process (unconstrained)*
1	2.08	4.35	–
2	6.25	4.35	–
3	16.66	21.74	1.06
4	25.00	15.22	11.53
5	20.83	30.43	26.47
6	18.75	10.87	29.59
7	6.25	13.04	19.22
8	4.17	0.00	8.48
9	0.00	0.00	2.80
10	0.00	0.00	0.81
	100.00	100.00	100.00
Mean contact	4.5833	4.478	6.000

England. It is evident that very regular patterns, like honeycombs, will have frequency distributions with a pronounced peak at a single value, while more complex patterns will show wider spreads around a central modal value.

The independent random process for generating polygonal areas outlined on pp. 47–8 and shown in Figure 3.8 gives the expected distribution given in column four of the table. The modal value is for six-sided hexagons. These values cannot be compared directly with those in columns 2 and 3 for two reasons. First, the method of defining the areas ensures that the minimum contact number must be three and, second, the procedure does not have edge-constraints, whereas both the USA and England have edges given partly by the surrounding oceans. In spite of these differences, it is apparent that these administrative areas have lower contact numbers than expected, implying a 'more regular than random' patterning.

Fragmentation and autocorrelation

The area patterns discussed in the previous section are usually of interest where they arise 'naturally' (see pp. 111–12). It is rather more frequent that data are available only for artificial or imposed areas, such as states, counties and parishes. These areal units are often unsuitable for the analyses we should like to conduct and only seldom is their spatial pattern of interest in itself. Instead, attention turns to the analysis of the patterns produced by the data which are related to those areas.

The simplest area map that can be produced from such data is a *binary map*, in which areas are coloured black (B) or white (W) according to the presence or absence of some nominally scaled phenomenon. The effect of this colouring will usually be to reveal a pattern of distinct regions formed of contiguous map areas shaded the same colour, as for example, the region of Ford voting west of the Mississippi–Missouri in the USA, revealed in Figure 5.8. Yet for a given number of areas coloured B and W the number of possible arrangements (i.e. maps) is given by

$$\binom{n}{k} = \frac{n!}{k!(n-k)!}$$

where n = number of areas and k = number of areas coloured B. Even with only sixteen cells, half of which are coloured black, this number of possible maps is

$$\binom{16}{8} = \frac{16!}{8!\ 8!} = 12{,}870$$

The number of *possible* arrangements of states voting for Ford rather than the actual pattern in Figure 5.8 is very much higher still. It follows that there is a need to develop measures that will enable us to discriminate among these possibilities. How likely is a particular pattern of the variable in question?

Examine Figure 5.8 again: it is apparent that there is a distinct tendency for states with a particular affiliation to group together. Most geographers would be prepared to use this as evidence on which to generalize about the geography of voting in the USA. Quantitatively it is an example of the phenomenon known as (positive) *spatial autocorrelation*, but qualitatively the idea has been recognized ever since the first thematic maps were drawn. Indeed, the often-quoted first law of geography that 'everything is related to everything else, but near things are more related than distant things' (Tobler, 1970) is simply a statement that spatial autocorrelation exists. Unfortunately, between this simple idea and the various statistical tests that have been proposed (Cliff and Ord, 1973) mystification seems often to set in, and one often hears quite eminent geographers claiming that they 'do not understand

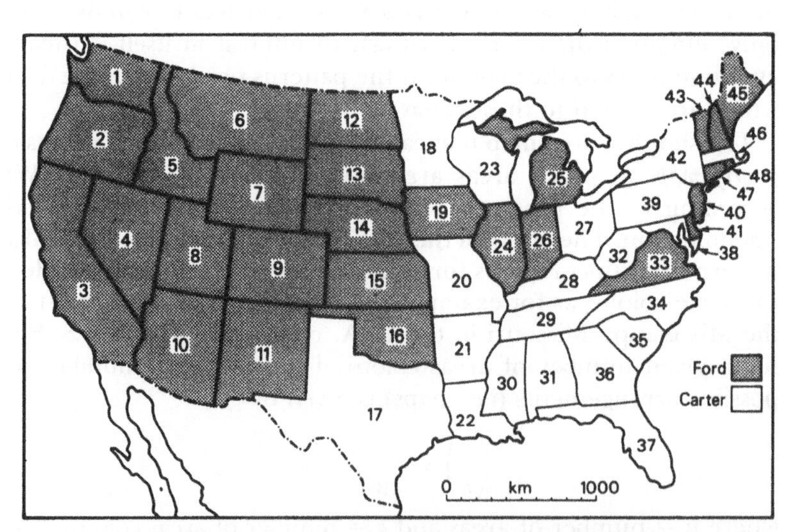

Figure 5.8 How the states in the USA voted in 1976.

spatial autocorrelation'. If this were really true, one should question their claim to be geographers at all! What they are saying is not that they do not understand the idea, but that they are not familiar with the mechanics of the statistical tests associated with it. This is a pity, there is nothing particularly mysterious about these either!

Consider a binary map of the type illustrated in the three examples in Figure 5.9. The critical step in the logic is to view these maps as the results of generating processes which determine how they are coloured B or W. The idea of process used here is rather more general than is often the case. Sometimes it will involve a true time sequence as in the diffusion of an innovation or the spread of an epidemic, but in the sense used here, it need not involve any passage of time at all. Simply, it is the imagined mechanism which sorts the areas – at a stroke, as it were – into the observed pattern. Such a sorting process can be deterministic, as in the instruction to 'colour every alternate square B', but in geography it is most often stochastic, involving a chance element. Just as was done for point and line data, the simplest such process we can postulate is the independent random, in which areas are independently coloured B or W according to some fixed probabilities. To simulate such a process with fixed probabilities of 0.5 of a B or W colouring, simply toss a coin once for each of the sixty-four areas shown in each of the maps in Figure 5.9, shading B for 'heads' and W for 'tails'. Even without performing this experiment, it can be seen that (a) and (c) in the figure are extremely unlikely realizations of this process, whereas results which look like the areas in (b) are much more probable. In Figure 5.9(a) areas coloured B are surrounded by similarly coloured areas and vice versa, making it almost certain that the colouring was not independent. Like values tend to be adjacent to each other, giving what is called *positive spatial autocorrelation*. Similar comments can be made about Figure 5.9(c) except that the tendency is for dissimilar colours to be adjacent, giving *negative spatial autocorrelation*. Figure 5.9(b) is a realization of a genuinely independent random process which exhibits no significant spatial autocorrelation.

There are a number of ways we could choose to measure this phenomenon of which the most obvious is to use the idea of a joins count. In this, the number of 'joins' of a particular type, black to black (BB), white to white (WW) and black to white or white to black (BW) are counted. In doing this, it is best to tabulate each

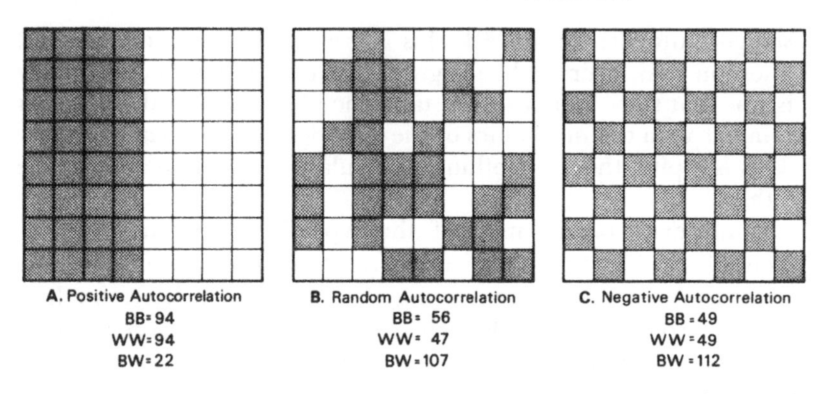

Figure 5.9 a, b and c The joins-count approach to spatial autocorrelation.

area and all its joins, so that each join will be counted twice, then divide the result by 2. Thus, the counts for the top row of Figure 5.9(b) are:

Cell Colour	BB	WW	BW	*Number of joins (j)*
1 W	0	2	1	3
2 W	0	2	3	5
3 B	3	0	2	5
4 W	0	2	3	5
5 W	0	3	2	5
6 W	0	4	1	5
7 W	0	3	2	5
8 B	0	0	3	3
	3	16	17	36

Notice that we have counted all joins, including those that only touch at the corners. The results of a complete count are given in Figure 5.9. There are 210 joins in all, but for the positive autocorrelation (Figure 5.9(a)) joins of like colour predominate (BB = WW = 94), and for negative autocorrelation on (Figure 5.9(c)) dissimilar joins are most common (BW = 112). Intuitively, it is understood that these counts act as indicators of the presence and type of autocorrelation, but can intuition be developed into a significance test? The approach adopted is exactly the same as that in the study of point and line patterns. First, an independent random process is postulated and probability theory used to predict what value of the joins counts would be expected in the long run. Remembering that the same process can give many realizations, it is also necessary to predict the variation, as measured by the standard deviation, of these expected values.

It can be shown that an independent random process gives the following results:

$$\text{expected number of BB joins } J_{BB} = kp^2$$
$$\text{expected number of WW joins } J_{WW} = kq^2$$
$$\text{expected number of BW joins } J_{BW} = 2kpq$$

where k = total number of joins on map; p = probability of an area being coded B; q = probability of an area being coded W. For all the patterns illustrated in Figure 5.9, $k = 210$ and $p = q = 0.5$, so that

$$J_{BB} = 210(0.5)^2 = 52.5$$
$$J_{WW} = 210(0.5)^2 = 52.5$$
$$J_{BW} = 2(210)(0.5)(0.5) = \underline{105.0}$$
$$210.0$$

Notice that these expected values must sum to k, the total number of joins, since $kp^2 + kq^2 + 2kpq = k(p + q)^2$ and by definition $p + q = 1$.

The formulae for the standard deviations of the expected independent random values are a little more difficult:

$$\text{standard deviation of } J_{BB} = [kp^2 + 2mp^3 - (k + 2m)p^4]^{0.5}$$
$$\text{standard deviation of } J_{WW} = [kq^2 + 2mq^3 - (k + 2m)q^4]^{0.5}$$
$$\text{standard deviation of } J_{BW} = [2(k + m)pq - 4(k + 2m)p^2q^2]^{0.5}$$

The only new term is m, defined as :

$$m = 0.5\sum_{i=1}^{n} j_i(j_i - 1)$$

where j_i = number of joins to the ith area.

For the patterns in Figure 5.9, m is found as follows. There are four corner areas each with three joins, twenty-four side areas with five joins each, and 36 interior areas with eight joins, hence:

$$\begin{aligned} m &= 0.5[4 \times (3)(3 - 1) + 24 \times (5)(5 - 1) + 36 \times (8)(8 - 1)] \\ &= 0.5(24 + 480 + 2016) \\ &= 1260 \end{aligned}$$

The standard deviations are thus:

standard deviation of J_{BB}

$$\begin{aligned} &= [210(0.5)^2 + 2(1260)(0.5)^3 - (210 + 2520)(0.5)^4]^{0.5} \\ &= [52.5 + 315 - 170.625]^{0.5} = 14.031 \end{aligned}$$

standard deviation of J_{WW}

$$\begin{aligned} &= [210(0.5)^2 + 2(1260)(0.5)^3 - (210 + 2520)(0.5)^4]^{0.5} \\ &= 14.031 \end{aligned}$$

standard deviation of J_{BW}

$$\begin{aligned} &= [2(210 + 1260)(0.5)(0.5) - 4(210 + 2(1260))(0.5)^2(0.5)^2]^{0.5} \\ &= [735 - 682.5]^{0.5} = 7.246 \end{aligned}$$

The probability of any one specified number of joins of a particular type, given the independent random process, can be found by calculating the standard normal deviate, or z-score, as :

$$z = \frac{\text{observed number of joins} - \text{expected number of joins}}{\text{standard deviation of expected values}}$$

and then referring this to tables of the normal distribution function. There are three such z-scores, giving three separate tests for each joins count. For example, the z–score for BB joins in Figure 5.9(a) is :

$$z_{BB} = (94 - 52.5) / 14.031 = 2.953$$

and the results for all nine tests of the three patterns are presented in Table 5.6. Since no direction, positive or negative, is specified for the autocorrelation, a two-tailed test is appropriate, and rejection of the null hypothesis that the values are drawn from a population whose true mean is 0 at the 95 per cent confidence level would need a z-score greater than $\pm$ 1.96 in magnitude. As can be seen, only in the case of pattern in (a) in the figure can the null hypothesis be rejected at this level.

Before going on to give an example of spatial autocorrelation tests using real-world data, five important points should be made. First, care should be taken to ensure that the direction of the

Table 5.6 z-scores for the joins counts (Figure 5.9)

	Pattern		
Type of join	A *positive autocorrelation*	B *independent random*	C *negative autocorrelation*
BB	*+2.953*	0.249	−0.249
WW	*+2.953*	−0.392	−0.249
BW	*−11.455*	0.276	0.966

Note: Italic numbers denote significance at 95 per cent.

autocorrelation is correctly interpreted; this is not simply positive or negative according to the sign of the relevant z-score. A moment's thought will indicate that the very high negative z_{BW} for the pattern in Figure 5.9(a) indicates positive autocorrelation. It tells us that there are many fewer BW joins than we expected, indicating a tendency for similarly coloured areas to group together.

Second, despite its very obvious regularity, the chequer-board pattern in Figure 5.9(c) is not picked out as significantly different from random. This arises because we chose to count all joins to areas, including the diagonal ones. By analogy with the moves the Queen can make in chess, this is called the 'Queen's case' and it results in nearly every area ending up with the random expectation of an equal number of neighbours of each colour. Had the analysis been restricted to the joins counts along only rows and columns – the 'Rook's case' – the result would have been very different as the analysis for BW joins confirms:

observed number of BW joins = 85

$$k = 85$$

$$m = 0.5[4 \times (2)(2 - 1) + 6(3)(3 - 1) + 36(4)(4 - 1)]$$

$$= 238$$

expected number of *BW* joins $J_{BW} = 2\,(85)(0.5)(0.5) = 42.5$
standard deviation of J_{BW}

$$= [2(85 + 238)(0.5)(0.5) - 4(85 + 476)(0.5)^2(0.5)^2]^{0.5}$$

$$= [161.5 - 140.25]^{0.5}$$

$$= 4.6098$$

$$z_{BW} = (85 - 42.5)/4.6098 = 9.220$$

The result is now indicative of the 'expected' highly significant, negative spatial autocorrelation. It also shows the importance of relating statistical hypothesis tests to the precise notion being tested and how more than one quantitative hypothesis can be embedded in an apparently simple, clearcut qualitative idea.

Third, these significance tests all make use of the tabulated normal distribution, assuming, in effect, that the joins-count statistics have such a normal distribution. In fact, as was shown by Cliff and Ord (1973), this is true only at the limit as the number of areas n approaches infinity. In practical terms, the normal approximation is reasonable only if n is moderately large, the probabilities p and q are not too near 0 and 1, and if the lattice of areas is not dominated by one or two very large areas. Unfortunately, it seems that for many systems of interest such as the counties in the UK or states in the USA n is small and the normal approximation poor. An alternative test, originally due to Dacey, which is appropriate for small n is evaluated in Cliff, Martin and Ord (1975). It uses a modified chi-square test of the observed and expected frequencies of number of areas with joins of a particular type.

Fourth, in calculating the expected numbers of joins, we assumed that the probabilities p and q were known in advance and *independently of the data used in the test.* This is called *free sampling* and is appropriate, for example, if the region being studied is set in a much larger region from which p and q can be found. However, in many geographical applications these probabilities will not be obtained in advance, but will have to be estimated from the same data used in the autocorrelation test as:

$$p = \frac{\text{number of black areas}}{\text{total number of areas}} = n_1/n$$

$$n_1 = \text{number of black areas}$$

$$n = \text{total number of areas}$$

$$q = (1 - p) = (n - n_1)/n$$

The necessity to estimate p and q in this way alters the expressions for both the expected joins, and their standard deviations because the effect is to constrain the colouring process such that, in total, there must be n_1 areas coloured B and $n_2 = n - n_l$ coloured W. This is called *non-free sampling.* If the probabilities are found in this way, the effect is to alter the expressions for the expected values. For example, those for BW joins become:

$$\text{expected number of BW joins } J_{BW} = 2kn_1n_2/\ n(n - 1)$$

standard deviation

$$= \left[J_{BW} + \frac{2mn_1n_2}{n(n-1)} + \frac{4\{k(k-1) - 2m\}n_1(n_1-1)n_2(n_2-1)}{n(n-1)(n-2)(n-3)} - J_{BW}^2 \right]^{0.5}$$

k and m have the same definition as for free sampling and the calculation of a z-score proceeds in exactly the same way as before.

Finally, we have dealt with only one test, applicable to binary data and autocorrelation between immediately adjacent areas. The restriction to binary data is not as serious as it appears. If the researcher is willing to accept the loss of some information, virtually all forms of data – ordinal, interval or ratio – can be coded B or W according to whether they are above or below the mean, or reduced to high (B) or low (W) categories. Alternatively, use can be made of other autocorrelation coefficients designed for these types of data, as outlined by Ebdon (1977). The second restriction – to immediate adjacencies – can be circumvented by examining joins at lagged distances and, apart from the awkwardness defining and counting joins, the test proceeds in exactly the same way.

An example

The joins-count test for spatial autocorrelation can be illustrated using data from Figure 5.8 on votes for Ford and Carter in the 1976 US presidential election. Is the evident concentration of the Carter vote a statistically significant one? Table 5.7 presents a list of the conterminous states mapped, their colour, contact number and counts for BB, WW and BW joins. Summing the columns yields the information that there are twenty-eight states coloured B and that the total number of joins is.

$$k = 0.5\sum_{i=1}^{n} j_i = 110$$

The observed number of joins are BB = 45, WW = 31 and BW = 34. For the test, we can make an independent estimate of the probabilities, p, q, of a Ford or Carter vote, using the numbers of representatives returned by each state. For Ford there are 255, and for Carter 285, giving probability estimates of:

$$p = 255 / 540 = 0.4722$$
$$q = 285 / 540 = 0.5278$$

Table 5.7 Joins count for the US electoral data of Figure 5.8.

			Joins						*Joins*		
State	B/W	j_i	BB	WW	BW	*State*	B/W	j_i	BB	WW	BW
Washington	B	3	3	0	0	Michigan	B	4	2	0	2
Oregon	B	4	4	0	0	Indiana	B	4	2	0	2
California	B	3	3	0	0	Ohio	W	5	0	3	2
Nevada	B	5	5	0	0	Kentucky	W	7	0	4	3
Idaho	B	6	6	0	0	Tennessee	W	8	0	7	1
Montana	B	4	4	0	0	Mississippi	W	4	0	4	0
Wyoming	B	7	7	0	0	Alabama	W	4	0	4	0
Utah	B	6	6	0	0	West Virginia	W	5	0	4	1
Colorado	B	7	7	0	0	North Carolina	W	4	0	3	1
Arizona	B	5	5	0	0	South Carolina	W	2	0	2	0
New Mexico	B	6	5	0	1	Georgia	W	5	0	5	0
North Dakota	B	3	2	0	1	Florida	W	2	0	2	0
South Dakota	B	6	5	0	1	Maryland	W	5	0	2	3
Nebraska	B	6	5	0	1	Pennsylvania	W	6	0	4	2
Kansas	B	4	3	0	1	New Jersey	B	4	1	0	3
Oklahoma	B	6	3	0	3	Delaware	B	3	1	0	2
Texas	W	4	0	2	2	New York	W	5	0	2	3
Minnesota	W	4	0	1	3	Vermont	B	3	1	0	2
Iowa	B	6	3	0	3	New Hampshire	B	3	2	0	1
Missouri	W	8	0	3	5	Maine	B	1	1	0	0
Arkansas	W	6	0	5	1	Massachusetts	W	5	0	1	4
Louisiana	W	3	0	3	0	Connecticut	B	3	1	0	2
Wisconsin	W	4	0	1	3	Rhode Island	B	2	1	0	1
Illinois	B	5	2	0	3	Virginia	B	5	0	0	5

To find the expected numbers, we also need to compute m, as in Table 5.8. The expected numbers of joins are thus:

$$J_{BB} = kp^2 = 110(0.4722)^2 = 24.529$$
$$J_{WW} = kq^2 = 110(0.5278)^2 = 30.640$$
$$J_{BW} = 2kpq = 2(110)(0.4722)(0.5278) = 54.830$$

Notice that these expected numbers sum to 110. Standard deviations are, for J_{BB}

$$= [kp^2 + 2mp^3 - (k + 2m)p^4]^{0.5}$$
$$= [110(0.4722)^2 + 2(453)(0.4722)^3 - (110 + 2(453))(0.4722)^4]^{0.5}$$
$$= [24.529 + 95.404 + 50.522]^{0.5}$$
$$= 8.331$$

Table 5.8 Calculation of the *m*-value for the spatial autocorrelation test

Contact number j_i	*Frequency of states* f_i	$f_i j_i$	$j(j_i - 1)f_i$
1	1	1	1 (1 − 0) 1 = 0
2	3	6	2 (2 − 1) 3 = 6
3	8	24	3 (3 − 1) 8 = 48
4	12	48	4 (4 − 1)12 = 144
5	10	50	5 (5 − 1)10 = 200
6	9	54	6 (6 − 1) 9 = 270
7	3	21	7 (7 − 1) 3 = 126
8	2	16	8 (8 − 1) 2 = 112
	48	220	906

$k = 220/2 = 110,\ m = 906/2 = 453$

for J_{WW}

$$= [kq^2 + 2mq^3 - (k + 2m)q^4]^{0.5}$$
$$= [110(0.5277)^2 + 2(453)(0.5277)^3 - (110 - 2(453))(0.5277)^4]^{0.5}$$
$$= [30.640 + 133.193 - 78.831]^{0.5}$$
$$= 9.220$$

for J_{BW}

$$= [2(k + m)pq - 4(k + 2m)p^2q^2]^{0.5}$$
$$= [2(110 + 453)(0.4722)(0.5277) - 4(110 + 2(453))(0.4722)^2(0.5277)^2]^{0.5}$$
$$= [280.631 - 252.434]^{0.5}$$
$$= 5.310$$

Hence:

$$z_{BB} = (45 - 24.529)/8.331 = 2.457$$
$$z_{WW} = (31 - 30.640)/9.220 = 0.039$$
$$z_{BW} = (34 - 54.830)/5.310 = -3.923$$

Of the three tests, those for BB and BW greatly exceed the critical value of 1.96 necessary for significance at the 95 per cent level, whereas the expected and observed values of WW joins are virtually identical and the resulting *z*-score is not statistically significantly different from zero. At first sight, these results may appear contradictory. We have three tests, two of which indicate that the null hypothesis should be rejected and one which argues that it should not. Remember that all a non-significant value indicates is that we have insufficient evidence to reject the null hypothesis, and *not* that we must therefore 'accept' it. On the other hand, the tests for BB and BW both indicate that, using these different pattern measures, there is evidence to reject the null hypothesis and accept the alternative that spatial autocorrelation is present in our map. This kind of result occurs very frequently in joins-count tests. Indeed, it is not uncommon for both BB and WW tests to be inconclusive, whereas for BW it is clear and unequivocal. Even though this looks like 'two-to-one against', the logical course of action is still to infer the presence of autocorrelation.

A second point to note about these results is that the signs of the *z*-scores are consistent with the presence of *positive* autocorrelation. There are more BB and fewer BW joins than expected, indicating a tendency for like-voting states to group together. In this case such a result would be expected from a simple visual analysis of the map and, as geographers, we should proceed to search for hypotheses to 'explain' it. Yet patterns which visually *seem* to be interpretable in the same way often turn out to have a spatial autocorrelation structure that is not significantly different from random, so that statistically there is no objective basis for the subsequent theorizing. There are firm grounds for suggesting that all spatial analysis should start with a test for spatial autocorrelation, simply in order to prevent this tendency to infer and interpret pattern where none exists.

A second reason for studying spatial autocorrelation is its influence on standard statistical tests of the sort to be found in any statistics textbook. Such tests almost invariably assume that observations are independent, yet if spatial autocorrelation is present, the assumption no longer holds and this will usually lead to serious inferential errors. (See, for example, Gould, 1970; Martin, 1974.) A third reason for studying autocorrelation is in modelling geographic phenomena where it appears either as a troublesome complication (Sheppard *et al.*, 1976), or more usefully, can be incorporated in mathematical descriptions of spatial structure and process (for a review, see Cliff and Ord, 1975). The spatial autocorrelation structure of a pattern has often been used as an index of the visual complexity of that pattern in research on the perception of map pattern (Olson, 1975). There is also a direct relationship between the value of the spatial autocorrelation coefficient and various other measures of map pattern, notably the *redundancy* and *complexity* as defined by measures based upon the information theory of communication (see Marchand, 1975; Gatrell, 1977).

CONCLUSIONS

Maps based on area data are probably the most common in the professional geographical literature, yet in many ways they are the most complex to draw and analyse. If insufficient care is taken in design, they can easily misrepresent any underlying real distribution and are often very difficult to read. Statistically, they can be analysed by postulating a process and then examining how likely a particular observed pattern of chorochromatic shades or choropleth values is as a realization of that process. Just as in point pattern analysis (Chapter 3) and network analysis (Chapter 4), the postulated process was an independent random one and departures from the mathematically derived expected values of various joins-count statistics were used to measure and construct a test for what is known as spatial autocorrelation. In conclusion, it is worth repeating that the idea of spatial autocorrelation is not a peripheral, rather abstract, idea that should be the concern solely of geographers with a 'quantitative' inclination. Rather it is a quantitative way of examining in a rigorous and objective fashion what has always been seen as a fundamental geographical idea – that of the patterns of similar or dissimilar areas on the surface of the earth.

RECOMMENDED READING

There is a very large and growing literature on the analysis of area patterns, of which the recommended reading cited below is a small but, hopefully, representative sample. The list is divided into three major areas, corresponding to sections in the text.

Cartographic considerations

Castner, H. W. and Robinson, A. H. (1969) *Dot area symbols in cartography: the influence of pattern on their perception,* ACSM Monographs in Cartography 1, Washington DC, Amer. Congr. Survey and Mapping.

Coppock, J. T. (1955) 'The relationship of farm and parish boundaries: a study in the use of agricultural statistics', *Geographical Studies* 2, 12–26.

Coppock, J. T. (1960) 'The parish as a geographical statistical unit', *Tijdschrift voor Economische en Sociale Geographie* 51, 317–26.

Dobson, M. W. (1973) 'Choropleth maps without class intervals? A comment', *Geographical Analysis* 5, 358–60.

Evans, I. S. (1977) 'The selection of class intervals', *Transactions, Inst. Brit. Geogr.*n.s. 2, 98–124.

Jenks, G. F. (1953) 'Pointillism as a cartographic technique', *Professional Geographer* 5, 4–6.

Jenks, G. F. and Caspall, F. C. (1971) 'Error on choroplethic maps: definition, measurement, reduction', *Annals, Assoc. Amer. Georgr.* 61, 217–44.

Jenks, G. F., and Knos, D. S. (1961) 'The use of shading patterns in graded series', *Annals, Assoc. Amer. Georgr.* 51, 316–41.

Muller, J. C. and Honsaker, J. L. (1978) 'Choropleth map production by facsimile', *Cartographic Journal* 15, 14–19.

Openshaw, S. (1977) 'A geographical solution to the scale and aggregation problems in region building, partitioning and spatial modelling', *Transactions, Inst. Brit. Geogr.* n.s. 2, 459–72.

Robinson, A. H. (1956) 'The necessity of weighting values in the correlation of areal data', *Annals, Assoc. Amer. Geogr.* 46, 233–6.

Robinson, A. H. (1967) 'The psychological aspects of colours in cartography', *International Yearbook of Cartography* 7, 50–61.

Robinson A. H., Lindberg, J. B. and Brinkman, L. W. (1961) 'A correlation and regression analysis applied to rural farm population densities', *Annals, Assoc. Amer. Geogr.* 51, 211–21.

Taylor, P. J. and Johnston, R. J. (1979) *The Geography of Elections,* Harmondsworth, Pelican Books.

Tobler, W. R. (1973) 'Choropleth maps without class intervals', *Geographical Analysis* 5, 26–8.

Williams, R. L. (1958) 'Map symbols: equal appearing intervals for printed screens', *Annals, Assoc. Amer. Geogr.* 48, 132–9.

Area measurement and shape

Baxter R. S. and Lloyd, N. (1972) 'Computers in land area measurement', *Technical Note* A4, Cambridge, LUBFS.

Berry, B. J. L. (1962) *Sampling, Coding and Storing Flood Plain data,* Washington DC, USDA, Farm Economics Division Handbook 237.

Boyce, R. and Clark, W. (1964) 'The concept of shape in geography', *Geographical Review* 54, 561–72.

Cerny, J. W. (1975) 'Sensitivity analysis of the Boyce–Clark shape index', *Canadian Cartographer* 12, 21–7.

Clark, W. and Gaile, G. L. (1975) 'The analysis and recognition of shapes', *Geografiska Annaler* 55B, 153–63.

Frolov, Y.S. (1975) 'Measuring the shape of geographical phenomena: a history of the issue', *Soviet Geography* 16, 676–87.

Frolov, Y. S. and Maling, D. H. (1969) 'The accuracy of area measurement by point counting techniques', *Cartographic Journal* 6, 21–35.

Gierhart, J. W. (1954) 'Evaluation of methods of area measurement', *Survey and Mapping* 14, 460–9.

Lee, D. R. and Sallee, G. T. (1970) 'A method of measuring shape', *Geographical Review* 60, 555–63.

Stoddart, D. R. (1965) 'The shape of atolls', *Marine Geology* 3, 369–83.

Switzer, P. (1976) 'Applications of random process models to the descriptions of spatial distributions of quantitative geological variables', in Merriam, D. F. (ed.) *Random Processes in Geology,* New York, Springer Verlag, Chapter 10, 124–34.

Taylor, P. J. (1977) *Quantitative Methods in Geography: An Introduction to Spatial Analysis,* Boston, Houghton Mifflin, especially 41–2 and 78–9.

Wood, W. F. (1954) 'The dot planimeter: a new way to measure area', *Professional Geographer* 6, 12–14.

Statistical, especially spatial autocorrelation

The literature on spatial autocorrelation is generally regarded as 'difficult', but in principle there is no reason why this should be so. As a basic introduction to the ideas with worked examples of three types of coefficient and a simplified notation, the following is highly recommended:

Ebdon, D. (1977) *Statistics in Geography: A Practical Approach,* Oxford, Blackwell.

Rather more difficult syntheses will be found in :

Cliff, A. D. and Ord, J. K. (1973) *Spatial Autocorrelation,* London, Pion.

Haggett, P., Cliff, A. D. and Frey, A. (1977) *Locational Methods,* London, Arnold, Chapter 11, 353–77.

More specific definitions and applications can be found in:

Cliff, A. D. and Ord, J. K. (1975) 'Model building and the analysis of spatial pattern in human geography', *Journal of the Royal Statistical Society* B 37, 297–348.

Cliff, A. D., Martin, R. L. and Ord, J. K. (1975) 'A test for spatial autocorrelation based upon a modified χ^2 statistic', *Transactions, Inst. Brit. Geogr.* 65, 109–29.

Gatrell, A. C. (1977) 'Complexity and redundancy in binary maps', *Geographical Analysis* 9, 29–41.

Geary, R. C. (1954) 'The contiguity ratio and statistical mapping' in Berry, B. J. L. and Marble, D. (eds) *Spatial Analysis: A Reader in Statistical Geography*, Englewood Cliffs, NJ, Prentice-Hall, 461–78.

Gould, P. (1970) 'Is *Statistix inferens* the geographical name for a wild goose?', *Economic Geography* 46, 439–48.

Marchand, B. (1975) 'On the information content of regional maps: the concept of geographical redundancy', *Economic Geography* 51, 117–27.

Martin, R. L. (1974) 'On autocorrelation, bias and the use of first spatial differences in regression analysis', *Area* 6, 185–94

Monmonier, M. W. (1974) 'Measures of pattern complexity of choroplethic maps', *American Cartographer* 1, 159–69.

Olson, J. (1975) 'Autocorrelation and visual map complexity', *Annals, Assoc. Amer. Geogr.* 65, 189–204.

Sheppard, E., Griffith, D. A. and Curry, L. (1976) 'A final comment on misspecification and autocorrelation in those gravity parameters', *Regional Studies* 10, 337–9.

Tobler, W. R. (1970) 'A computer movie simulating urban growth in the Detroit region', *Economic Geography* 46, 234–40.

WORKSHEET

Figure 5.10 shows the ward boundaries for Sheffield, England, in 1971, and Table 5.9 gives population data for the same twenty-seven wards. The numbers on the map correspond to those in the table:

(1) Produce a choropleth map to show the percentage of population in each ward which originates from the New Commonwealth. Attempt to compare this distribution with that for Leicester (Figure 5.2), noting how difficult this is and how dependent on the choice of classes.

(2) In successive steps, reduce your choropleth first to a four-colour map, ordinally scaled from areas of 'high' to 'low' immigration and then to a two-colour map. Use

Continued on next page

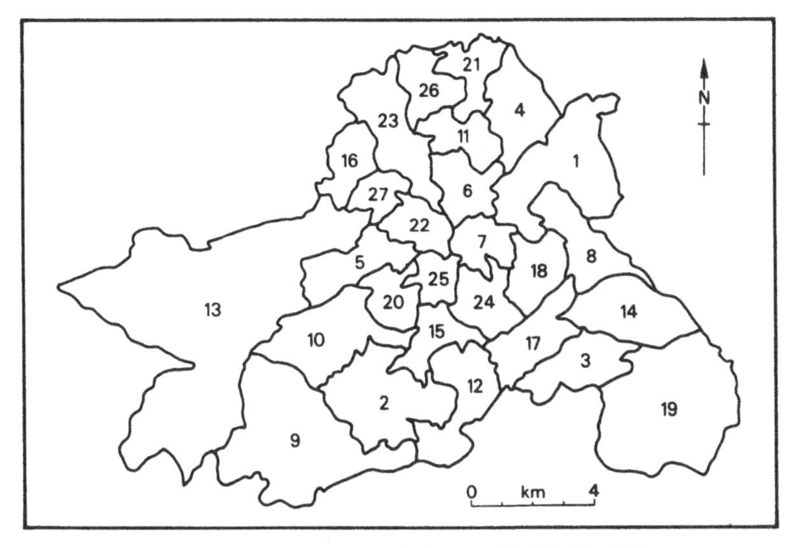

Figure 5.10 Ward boundaries in Sheffield, 1971.

either the quartiles of the distribution, or the mean and standard deviation.

(3) Find the percentage error of your map in exercise 1, and compare it with that for the simpler maps produced in exercise 2. For some comparative examples, see Coates, B.E. *et al.* (1974) *Census Atlas of South Yorkshire*, Sheffield, Department of Geography, University of Sheffield.

(4) As indicated on p. 114 use of colour on maps of any kind is complex. A very useful exercise is to list and describe the colours used on an OS 1 : 63,360 sheet and compare this with the more modern 1 : 50,000, noting how much of the obvious improvement in clarity is associated with changes in the colouring used. A suitable pair of maps from North America are the 1947 and 1956 editions of the 1 : 250,000 sheet NJ 172, '*Clarksburg*', produced by the US Department of the Interior, Geological Survey. Even though separated by less than a decade from its predecessor, the more recent sheet shows an enormous increase in clarity.

Continued on next page

Table 5.9 Population and numbers from the New Commonwealth, Sheffield, 1971

Number on map	*Ward name*	*Total population*	*Number from New Commonwealth*
1	Attercliffe	17,013	994
2	Beauchief	19,705	51
3	Birley	22,639	20
4	Brightside	18,529	180
5	Broomhill	17,300	326
6	Burngreave	17,967	590
7	Castle	17,596	192
8	Darnall	19,131	318
9	Dore	18,815	45
10	Ecclesall	19,764	155
11	Firth Park	18,433	179
12	Gleadless	23,360	29
12	Hallam	20,273	81
14	Handsworth	18,736	11
15	Heeley	21,623	287
16	Hillsborough	19,404	28
17	Intake	22,407	24
18	Manor	19,883	46
19	Mosborough	13,226	12
20	Nether Edge	18,586	467
21	Nether Shire	16,543	18
22	Netherthorpe	15,283	195
23	Owlerton	19,998	48
24	Park	21,668	59
25	Sharrow	15,605	406
26	Southley Green	18,875	20
27	Walkley	17,295	56

(5) Use the pattern of areas of Figure 5.2 to produce an adjacencies matrix, recording a 1 to indicate wards in contact. What is the average contact number, and how does it compare with those given for the US states and UK counties?

(6) Figure 5.11 shows a two-colour map of the political complexion of the English counties in 1973. Visually

Continued on next page

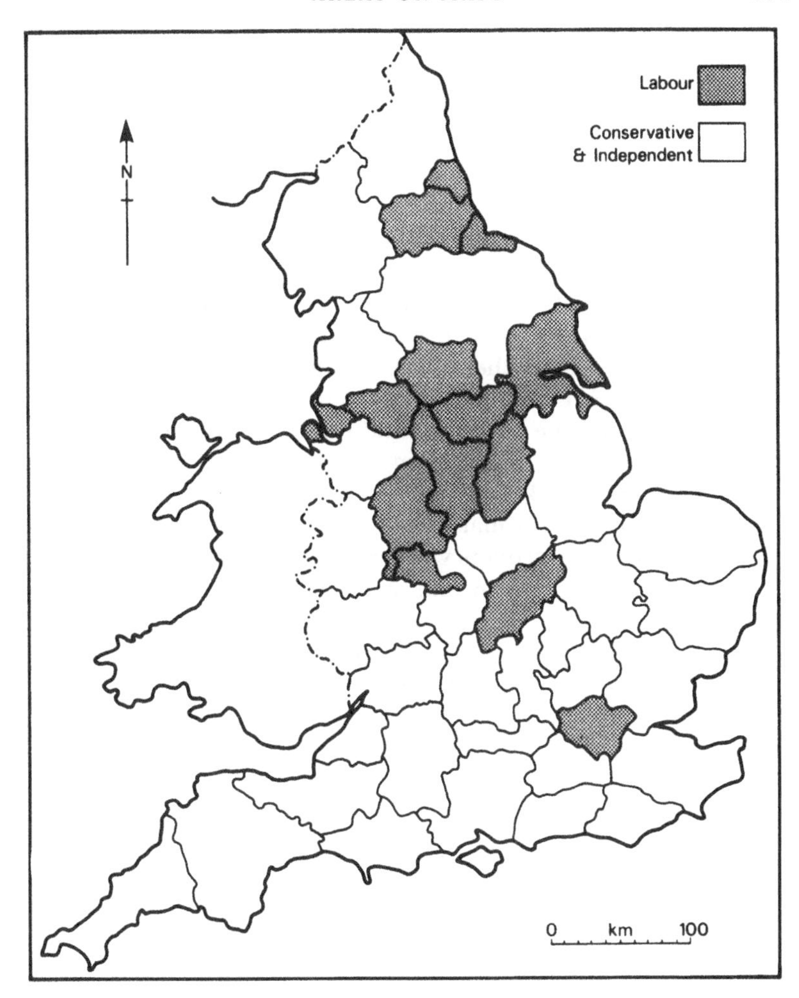

Figure 5.11 The political complexion of the English counties, 1973 (local elections).

there appears to be a concentration of Labour counties in the north and north-east, a pattern which can be explained fairly readily, but does a joins-count test allow such a generalization to be made? (Use the BW joins, non-free sampling.)

(7) An in-class sampling distribution experiment: for an introduction to the philosophy of this approach see Silk,

Continued on next page

J. (1979) 'Use of classroom experiments and the computer to illustrate statistical concepts', *Journal of Geography in Higher Education* 3, (1), 13–25. We stated on pp. 136–40 results for the expected average and standard deviation of the joins counts resulting from an independent random process, and asserted without proof, that these counts have a normal distribution when the number of areas is very large. It is highly instructive to compare these theoretical results with the results of a simulation exercise. Each student in a class of, say, thirty (the more, the better) should randomly colour cells of an 8 × 8 chequer-board, using a coin-toss to decide on each cell's colour, B or W, and then count the number of BW joins in the normal way. Pool these individual results, forming a frequency distribution of the counts and finding their mean and standard deviation. Now compare this simulated sampling distribution with the theoretically expected one. How close are the mean and standard deviation? How good is the approximation to normality with $n = 64$? If time allows, it is even more instructive to repeat the experiment using a 4 × 4 and then a 2 × 2 board.

CHAPTER SIX

·SURFACES ON MAPS·

INTRODUCTION

The final type of map recognized in Chapter 2 is that involving surfaces enclosing volumes, that is, maps with the spatial dimension L^3. A simple example is a map of a mountain peak (see Figure 6.1) which shows the altitude of the ground *surface*, but in so doing also defines the three-dimensional volume of the mountain. Although Chapter 2 allowed for the possibility of volumes defined by nominal (mountain/plain), ordinal (high ground/medium ground/ low ground) and interval or ratio data (altitude), in practice surface maps almost invariably show interval or ratio data, and techniques for the display of nominal and ordinal surfaces have already been discussed in Chapter 5.

To introduce surface mapping, we shall make use of a mathematical concept, the *scalar field.* To a mathematician, a scalar is any quantity characterized only by its magnitude or amount and independent of any spatial co-ordinates in which it is measured. An example of a scalar is temperature; one number gives its magnitude and this remains the same irrespective of how we transform its spatial position using different map projections. A scalar field is any graph showing the value of the scalar as a function of its position as, for example, the contour map shown in Figure 6.1(b). Scalar fields can be represented mathematically by a very general equation:

$$z = \mathrm{f}(x, y)$$

where z = scalar magnitude; x, y = spatial co-ordinates; and f denotes 'some function'.

For the mountain all this says is that the altitude (z) is related to position (x, y), but two critical assumptions have already been made. The first is that of continuity, that is a z-value exists (or can be imagined to exist) everywhere on the surface, and that there are no sharp discontinuities. On the mountain this implies that there are

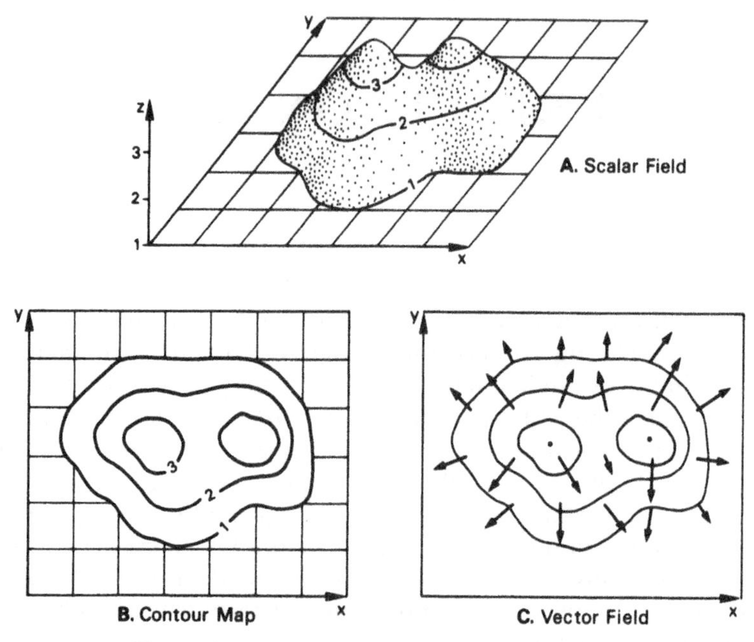

Figure 6.1 a, b and c The concept of a field.

no bottomless pits or vertical cliffs. Second, it is assumed that the field is single-valued, that is only one value of z is present at each location (x, y). On the mountain this implies that there are no overhangs. In physics, scalar fields are usually simple, and so are capable of being represented by straightforward mathematical functions, but those fields of interest to the geographer, as for example rainfall, temperature, population potential, travel time, economic rent, and so on, are usually much more complex. Geographers habitually map, describe and analyse fields of numbers, so that in a very special sense geography is 'field science', but this does not necessarily mean 'fieldwork'! What is often forgotten is that the study of fields is a valuable and practical tool for understanding many of the principles of geometry, mechanics and magnetism, and has a very large and valuable literature (see McQuistan, 1965, for an introduction).

THEORY AND PRACTICE IN SURFACE MAPPING

Scalar fields can be mapped in a number of ways, from relatively simple point representations through to complex isolines.

Point height data

The simplest representation is to plot the surface height at a series of points, as shown in Figure 6.2(a). The points chosen could be significant ones on the surface, such as mountain peaks and valley bottoms, or they could be the result of a spatially random sampling. Ordinary topographic maps often show this kind of point-height data in a number of ways, the most common of which is the *spot height*, shown as a small point symbol (usually a dot or small open circle), whose z-value is written alongside. Both the height (z) and the location (x, y) are known very accurately, but there is no mark on the ground as well as on the map. In contrast, *bench marks* and *triangulation points* not only are represented on the map, but also are found marked on the ground as well. They enable further, more detailed field survey to proceed, using them as definite, known points. Indeed, the term 'bench mark' is becoming widely used to mean something 'known definitely'.

Point-height information has the advantages of absolute accuracy and honesty. It provides only those data which are known and does not offer the map user any interpretation of these data. The major disadvantage is, of course, that no real visual impression of the shape of the field is given.

Hachures

A second way of representing a scalar field concentrates attention on its shape as deduced from its gradient, and introduces the idea of a *vector field.* It will be remembered that a scalar is a quantity which is characterized solely by its magnitude. In contrast a *vector* quantity needs both a magnitude and a direction for its specification, as for example the wind, migration flows, the flow of water in a river, and so on. Vectors will change, if we transform the space in which they are mapped. Just as it is possible to examine scalar fields in a region filled with numbers, so we can have vector fields, that is fields of vectors whose magnitudes and directions change from point to point in space. A good example would be the vector field of winds used in meteorology. Such a field can be represented as an assembly of vectors, shown as located arrow symbols whose length is proportional to the vector magnitude and whose direction and sense (up/down, to/fro) is given by the arrow's direction.

An extremely important point to note is that *every scalar field has a*

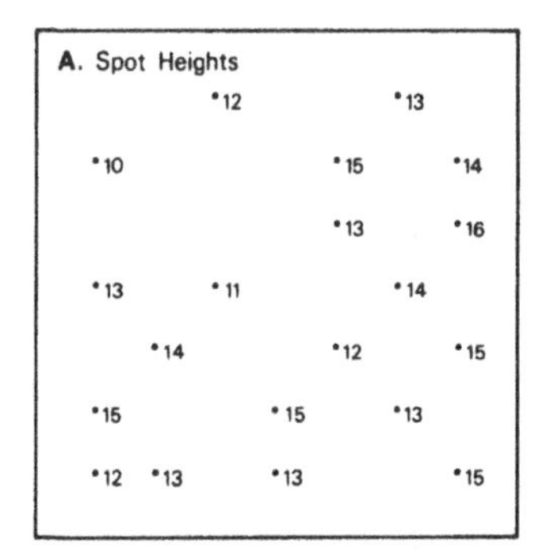

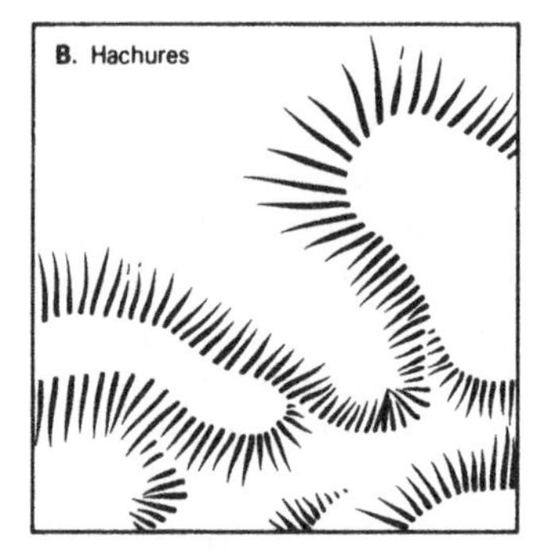

Figure 6.2 a and b Spot heights and hachures.

vector field associated with it, given by its gradient. Consider Figure 6.1(a) and (c). What would happen if we were to release a ball at every point on the mountain? Clearly, the ball would roll down the maximum slope, accelerating as it did by an amount related to the slope steepness. We can plot this magnitude and direction as a single downslope arrow. Repeating the exercise at a number of points, gives the vector field shown in Figure 6.1(c); we have, in effect, created a map of the maximum valley slopes or gradients. In this example of a mountainside, the associated vector field has an obvious and direct physical interpretation, but the vector fields of virtually all scalar fields have similar, useful substantive meanings. Thus in meteorology the vector field of the scalar temperature gives the direction of heat flow, the normal to the geostrophic wind is given by the vector field of the scalar pressure, and so on. These examples are from physical geography, but in human geography, Tobler (1976) has mapped 'migration winds' (vectors) from a scalar field of population.

A map of the vector field will always give a moderately good visualization of the surface relief and is the essence of the method of hachures used in early relief maps. A *hachure* is a line of varying length, width, or spacing drawn down the gradients of the scalar field, and the effect is to produce what is often called a 'hairy caterpillar' map (Figure 6.2(b)). The gradient direction is shown by the hachure direction and the magnitude by variations in the overall lightness or darkness of the result. A darker impression, indicating steep gradients, can be given using longer, thicker or more closely spaced lines. In themselves, hachures are a very inadequate way of representing relief and give very little useable quantitative information, but in the absence of other information, they do give a useful preliminary impression of the landform. As will be discovered if you try to draw any, they require considerable artistic ability. Usually, the effect is to give a very confused and cluttered map.

Isolines

Spot heights give accurate point information, hachures give a good visual impression of slope. If we could combine the advantages of both in a single form, then we would have a very good way of representing scalar fields. The *isoline* (also *isarithm*), of which the familiar relief contour is the best-known example, manages to achieve this combination. In isoline mapping we make an imaginary connection of all points of equal magnitude (z) to form a three-dimensional curve as shown in Figure 6.1(a). These three-dimensional curves are then projected on to a two-dimensional surface, usually, but not necessarily, the (x, y) plane at some assumed base height or datum. The resulting two-dimensional curves are called isolines, or lines of equal value (Figure 6.1(b)). Notice four things about this representation. First, the isolines show both the absolute magnitude of the scalar, and by their spacing, also provide information on the surface gradient. Second, it relates all the surface heights to a single, fixed datum plane from which all the z-values are measured. Most relief maps use the mean sea level as datum, but equally satisfactory datum planes in other applications would be 0° K, 0 mm rainfall, and so on. It is also possible to map depths below some datum as, for example, in isobases of sea or lake depth. Isolines are often given special names to indicate the nature of the scalar being mapped; thus isotherms (temperature), isovels (speed), isohyets (rainfall), isonoets (intelligence), isohels (sunshine), isobar (pressure), isochrone (time/distance), isophers (freight rate), isotachs (wind speed), and so on. Third, the isoline plot is the view we should get from an infinite distance above the scalar field, that is it is an orthogonal view, and finally, in order to draw isolines, we need to know a great deal about the variation in height of the original field.

In geography, two quite distinct types of isoline have been used, and it is well to be aware of the difference between them. The first are conventional isolines of the type discussed above, in which the surface being represented exists (or can be imagined to exist) at all points. Very different are the second type of isoline maps, which display spatial variation in derived quantities that are themselves related to some area, such as the population density (number per square kilometre). In the literature, isoline representations of data of this second type are called *isopleths*; they have been much used in agricultural and population studies. An isopleth map of population density, for example, can be produced by calculating the density of people over each of a number of census divisions, assigning these densities to some control points within each and every area, then threading isolines through the resulting field of numbers. This technique has a number of problems. First, it is apparent that, unlike normal isolines, the spaces between the isopleths do not have any values related to them. Second, the actual form of the isolines used will depend upon the shape of the areal units used and the locations of the control points. During the 1920s isopleth maps were rejected out of hand, but in the 1950s a number of papers, notably by Mackay (1951), Schmid and MacCannell (1955) and Porter (1958), developed rules for 'putting the isopleth in its place' and, more recently still, the development of automatic, computer cartography (Nordbeck and Rystedt, 1970; Tobler and Lau, 1978) has tended to encourage their use. While agreeing that isopleths can give a useful visual summary of a distribution, the author is inclined to argue that the best scientific solution to the isopleth problem is to refrain from drawing them! In the last analysis, area-based data should be displayed and analysed by area-based methods and not misrepresented as surface data.

The representation of isolines on a map is a tedious, but not too difficult task. Usually they are depicted as thin, continuous lines of appropriate colour, broken in places to allow value labelling, as in Figure 6.1(b), and the cartographic considerations are much the same as for any line information (see pp. 70–2). To aid interpretation, the fourth or fifth line is often thickened to act as a marker. The most important factors governing the final appearance of an isoline map are the number of isolines and the interval used. A large number of isolines will give a clear picture of the relief but require a great deal of height information and can obscure other detail; a small number require less data and will not mask other detail but will give a poor-relief picture. Deciding on the interval, is much the same as deciding on class intervals for choropleth maps, discussed in Chapter 5 (pp. 121–3), except that there are strong reasons for always using equal-interval height classes. If we

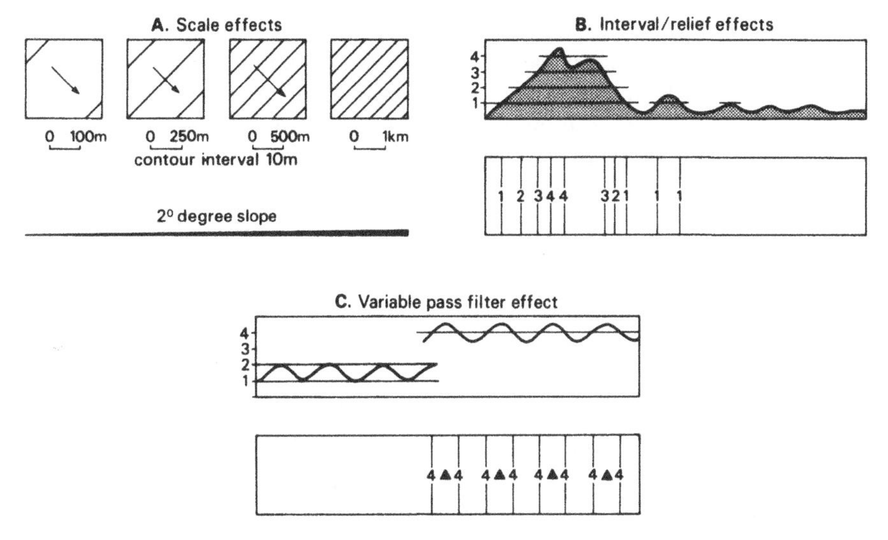

Figure 6.3 a, b and c Some isolining problems.

wish our isolines to show surface slope by their proximity to each other, there must be a standard, equal interval. Topographic maps have been published which do use a different contour interval above and below a certain altitude, but these should be interpreted with care.

The decision on what interval to use is not simple, depending on the scale and use of the map and, unfortunately, the nature of the surface being mapped. The effect of scale is obvious, but often forgotten. As Figure 6.3(a) shows, the same two-degree slope and 10-m contour interval look very different when shown on maps of differing scale. The effects of the relief itself are complex and difficult to generalize about. For example, the UK Ordnance Survey 1:50,000 sheets have an interval of either 15.24 m (50 ft), or on some more recent sheets, 10 m (30.48 ft). The 10-m interval is very much a compromise: in very mountainous areas, such as the Scottish Highlands, it produces a clear representation of the relief and only gives excessively crowded contours on the steepest slopes. Yet the same 10-m interval used in the flattest areas of the English lowlands may fail to pick up significant features. The problem is shown schematically in Figure 6.3(b). A final difficulty is illustrated in Figure 6.3(c), where the isolines act as variable filters according to the relationship of the feature height, altitude and contour value. The idealized lowland 'drumlin' field to the left of the inset has drumlins of exactly 0.99 units height, yet these fail to show up on the contour map below, whereas the same height features elsewhere which are cut by a contour plane (to the right of the figure), form prominent map features. Familiarity with contour maps often leads us to forget these unfortunate properties of isolines; and further, advanced geography students often forget how difficult contour maps are to interpret, if one is not familiar with them. To the trained eye they give a good impression of relief, but for the less skilled user it is often necessary to supplement them with other methods.

Other enhancing techniques

The three techniques discussed so far – spot heights, hachures and isolines – can be supplemented by a variety of other methods aimed at improving the overall visual impression of relief. Usually, these additional techniques supplement a contour or spot-height relief map, but there is no reason why they should not be used on their own, or to enhance the appearance of an isoline map of a scalar other than altitude. One such technique is that of *layer colouring*, sometimes known as *step surface shading*. This divides the height range into a series of bands, then shades the areas in each band an appropriate colour. When used to enhance isolines of, say, temperature, precipitation, or population potential, a graded series of intensities of a single colour can be used, as long as care is taken to ensure that the highest bands have the most intense shade and that there are no inversions in the sequence. Topographic maps are often layer-coloured, using a sequence of different colours usually from green, for the lowland, through yellow and brown to blues and whites, for high mountain peaks. Again, care should be taken to ensure that the intensity increases with altitude and that the colours used at each step conform roughly to popular associations. Whereas white peaks may be appropriate in the French Alps, they would certainly give a false impression of great altitude in the English Peak District.

It will be observed that layer colouring presupposes that some contours have already been plotted and that its major drawback is a tendency to generate an impression of a stepped landscape. This stepping can be made less obvious by *vignetting*, that is merging the colours across a contour boundary to give a gradual colour change; however, this requires great skill in both the selection and application of the colours. One example is given on the 1 : 63,360

Ordnance Survey sheet 'Tourist Map of the Cairngorms', published in the early 1960s.

A second supplementary technique which is often used on relief maps, but seldom on other isolines, is *hill shading*, which is a slightly more complex and updated version of the hachure. In it a continuous variation from light to dark wash is used to indicate the surface slope, the darker shade corresponding to the steeper slopes. Initially, the method became popular as lithography was developed, but nowadays it uses a halftone plate prepared either from a photograph of a relief model, or a monochrome wash. The relief is realized as if it were seen under illumination from a very distant, hence parallel, light source, the precise effect varying according to the source position. If this is vertically above, no shadows are created, but the amount of illumination varies in relation to slope angle. The steeper the slope, the less is the illumination, and this can be used to give the required shading. Alternatively, many maps use an assumed oblique source in the north-west, which gives light north-west facing and dark south-east facing slopes. This gives an excellent visual impression of relief but is not without its problems. Whether or not a slope is in shadow, depends as much on its azimuth as on it angle of slope. For example, a very gentle south-east facing slope may be illuminated as if it were a steep north-west facing one; whereas a steep south-east facing slope could be in shadow even though steeper slopes elsewhere are illuminated. At these extremes, the effect is not very misleading, but for slopes facing north-east or south-west the detailed impression can be very confusing indeed. A further difficulty occurs if the map is orientated upside-down, that is with north at the bottom and the shadows running away from the user. For reasons which are not fully understood, the effect is often a startling relief inversion; valleys become hills and *vice versa*. The problems are often circumvented by allowing the light source to 'float' anywhere from north-west to north-east and using one's artistic judgement to decide which slopes to shade. Finally, the advent of computer cartography has enabled a wide variety of analytical methods of displaying or enhancing the impression of relief but, as yet, none seems to have gained wide acceptance.

DESCRIBING SURFACES

Scalar fields are of interest in most branches of geography, however, physical geographers have tended to develop most of the available methods for summarizing and describing them. Geomorphologists interested in land form have analysed the fields of altitude, deriving useful information from properties such as the average altitude, the frequency distribution of altitude and the ground shape as expressed by its slope. Similarly, climatologists analyse the field of atmospheric pressure to derive the predicted geostrophic wind, while hydrologists find the total basin precipitation from a field of precipitation depths. Often the same, or similar, methods have been developed and named independently, and there is an enormous variety of possible ways of describing surfaces. In this account, we shall deal with a selection of these thought to be representative, the relative relief and average height, the area/height relationship, profiles and, finally, a way of calculating and mapping the gradient of the field.

Summary height measures

Perhaps the simplest measure used is the *relative relief*, which is the height range from lowest to highest point over some clearly specified area. A map of relative relief over a network of small grid squares gives a useful indication of the roughness of the surface and has the merit of being very easy to compute. The *average height* of a surface is rather more difficult. If an accurate isoline map is available, the simplest way is to calculate the average height from the heights of the midpoints between the contours, weighted by the proportion of area enclosed within the same contours, that is, from

$$z = \frac{1}{100} \sum_{i=1}^{k} \begin{matrix}\text{surface}\\ \text{height at contour } z_k\\ \text{midpoint}\end{matrix} \times \begin{matrix}\text{percentage total}\\ \text{area enclosed within}\\ \text{contours considered}\end{matrix}$$

Table 6.1 shows the calculations necessary to find the average height of the land surface in the contour map (Figure 6.4). As can be seen, the average height is almost 119 m, the only difficulty in the calculation being to determine the areas enclosed within each successive pair of contours. Any of the methods suggested on pp. 126–8 may be used, but for relatively complex contour patterns these will be found to be very laborious. A more rapid alternative, one which is particularly suitable, is to lay a network of equally spaced transect lines across the area and to measure the total length of the intercepts with each contour band. The percentage areas required are simply the equivalent percentages of the total line

Table 6.1 Calculation of average height, using area weighting

Contour band (*m*)	*Area* %	*Mid-point*	*Product*
less than 90	1.9	85	161.5
90–100	21.6	95	2052.0
100–110	10.7	105	1123.5
110–120	12.7	115	1460.5
120–130	17.6	125	2200.0
130–140	27.0	135	3645.0
140–150	6.8	145	986.0
150–160	1.2	155	186.0
greater than 160	0.5	165	82.5
	100.0		11897.0
	Average height = 118.97 m		

length. Once an average surface height is computed, then the enclosed *volume* is this average multiplied by the total mapped area. Although relative relief, average height and surface volume can be useful, they have the disadvantage that they do not give any real indication of surface shape and rely very heavily on the isolines being correctly located.

The area/height relationship

A rather more useful way of describing a surface is to examine the areal frequency of particular heights in some form of area/height histogram. Plots of the proportion of area at differing heights have been used a great deal in geomorphology in attempts to detect the existence of flat planation surfaces (for a review of the numerous available techniques see Clark and Orrell, 1958; Dury, 1972). Figure 6.4 shows an area/height histogram derived from the contour map and, alongside, a cumulative graph (sometimes called a *hypsometric curve*) showing the percentage of the land above or below any given altitude. Both show what might be held to be important flat surfaces at 90–100 and 130–140 m. When used in land-form analysis, these displays have many obvious deficiencies. They greatly rely on the initial accuracy of the contours and may well include flat features which, in the field, would have been excluded from consideration; they are also very laborious to compile.

A less laborious alternative is to derive the same end-product

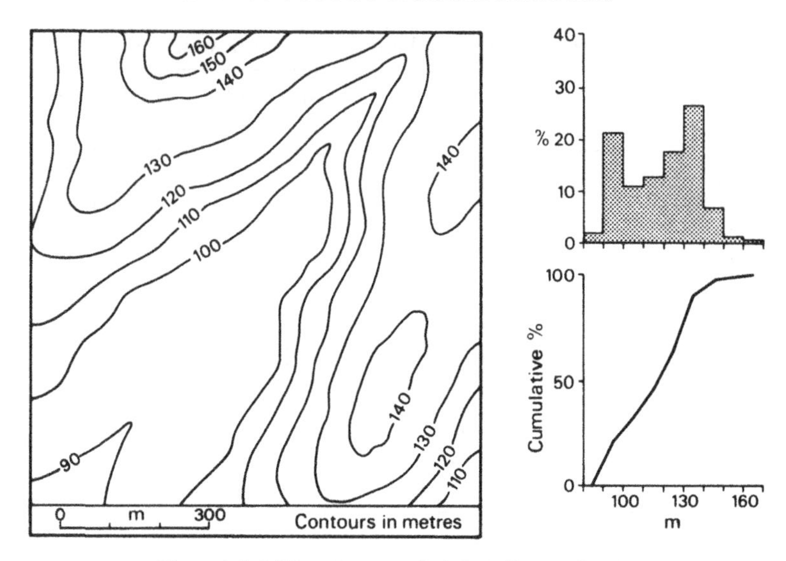

Figure 6.4 The average height of a surface.

from a histogram of spot-height frequency in differing height ranges. Indeed, almost all of the area/height measures which have been proposed can be replaced, or more readily computed, using as a starting-point not an isoline map, but an *altitude matrix* of the sort shown in Figure 6.5. This is a matrix representation of the map shown in Figure 6.4, derived by superimposing a fine, regular 10 × 10 mesh of spot heights and interpolating surface height at each of the 100 points. Nowadays, this type of matrix map is commonly used. First, they are often derived directly from photogrammetric heighting using aerial photographs or from other forms of remote sensing and, second, are very readily stored and processed by computer. As an example, the average height of our surface can simply be found without recourse to area measurement as the arithmetic mean of the 100 points which comes to 118.5 m, close to the value derived from the much more laborious method. Similarly, an altitude frequency histogram from these points gives a good idea of the area/height relationship and is very readily computed.

Profiles

Even with an area/height histogram, it is often difficult to visualize the height dimension of a surface. To aid visualization, and

•115	•130	•144	•162	•160	•141	•133	•130	•130	•132
•114	•130	•142	•145	•139	•131	•121	•115	•131	•135
•116	•130	•135	•131	•130	•120	•100	•125	•133	•141
•118	•131	•130	•120	•112	•100	•100	•130	•140	•140
•120	•124	•119	•108	•100	•95	•100	•130	•139	•136
•115	•111	•100	•95	•93	•99	•120	•132	•136	•132
•107	•100	•96	•93	•93	•100	•125	•140	•139	•129
•100	•96	•91	•93	•96	•110	•131	•145	•131	•117
•92	•91	•90	•94	•100	•120	•134	•139	•125	•109
•85	•85	•91	•100	•108	•118	•129	•125	•115	•102

Figure 6.5 The altitude matrix from Figure 6.4.

sometimes as an analytical tool, use can be made of *profiles* drawn across it. Often, the line of profile will be a straight line, but equally it can be an irregular line such as a river bed (giving a river-long profile) or a road. Drawing height profiles is not difficult, but care should be taken not to exaggerate unduly the vertical dimension. Unless the horizontal scale is large and the relative relief great, profiles will need to emphasize the vertical scale by a factor known as the *vertical exaggeration*. This is the ratio of horizontal to vertical scale expressed as the number of times the vertical has been stretched. Thus, a profile drawn to a horizontal scale of 1:50,000 which uses 10 mm to show 100 m in the vertical has a vertical exaggeration of 50,000/10,000 = 5.

A single profile gives only a limited visual impression of surface form, thus over the years several ways have been developed of showing more than one profile on the same plot. Profiles may be *superimposed* or, more usefully still, *composite*, in which the series are plotted as if viewed in the horizontal plane of the summit levels. Nowadays, it is a relatively easy matter to programme a computer to draw profiles, offsetting them to give an *isometric plot*, or adding perspective to give a *block diagram*. These plots all have their uses, but fundamentally all that they do is restate one graph (the isoline map) as another (the diagram).

Slopes and gradients

As we have seen, the real advantage of the isoline lies in its ability, through the spacing, to give a visual impression of surface slope, and slope is perhaps the most important property of any scalar field. The *slope* of a surface is the rate of change of height with horizontal distance. It may be expressed as an angle from the horizontal, as the rate of change itself (metre per metre), or as a ratio of height gained to distance travel (1-in-5, etc.). The calculation of slope is illustrated in Figure 6.6. Suppose that you intend to walk from the point A on Figure 6.6(b) to the top of the hill at B. The vertical interval (VI) separating these points is 610 – 200 = 410 m and the horizontal equivalent (HE) of the path distance measured at 2.8 km. As Figure 6.6(a) shows, the slope can be found from simple trigonometry as:

$$\begin{aligned}\tan\theta &= \text{opposite side/adjacent side}\\ &= 410/2800 = 0.1464\end{aligned}$$

In calculating this tangent, care must be taken to ensure that both numerator and denominator are in the same length units, in our

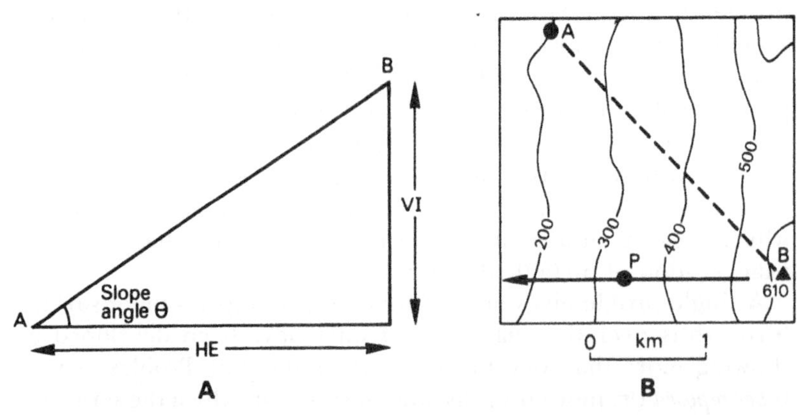

Figure 6.6 a and b The calculation of slope and gradient.

case metres. Looking it up in tables, shows that it corresponds to a slope angle of 8.35°, which is pretty steep! Notice several important things about slope. First, we derive it as a length divided by a length, so that dimensionally it is $LL^{-1} = L^{\circ}$, a dimensionless number whose value is independent of the units of measurement. Second, it is directional, calculated along the specified direction of the track and, third, it has a sense whether up or down.

Consider now the point P, also shown in Figure 6.6. Can we meaningfully measure the slope at this point? Clearly, to do this requires that we specify in advance a direction in which we want our slope to be measured. Any direction could be chosen, but three particularly useful ones are parallel to the y axis (towards north), parallel to x (towards east) and, most useful of all, down the maximum slope at the point which gives the *gradient*. It is this maximum slope that we met in the discussion of vector fields and hachures, and it is this gradient to which geomorphologists refer when they talk of slope angles.

In trying to measure and map gradients, geographers have developed a number of arbitrary and usually rather unsatisfactory methods of which that by Raisz and Henry (1937) is best known. Their method involves dividing a topographic map into a series of provinces by visual inspection of the contour spacing, then calculating the average gradient in each. The calculations may be speeded up, using templates drawn for a given contour interval and map scale against which the observed contours are matched, as developed by Thrower and Cooke (1968).

The literature on slope mapping is voluminous, but what seems only rarely to have been realized is that calculating the gradient of a scalar field is a perfectly standard operation in mathematics and physics, so much so, in fact, that it is often denoted by a special operator ∇, called 'del' (sometimes 'grad', more rarely, 'nabla'). We can create truly objective and continuous slope maps that make use of this concept as follows.

The starting-point is an altitude matrix, such as that of Figure 6.5. The chosen grid length is 100 m, and will control the overall accuracy of the final maps. Examine Figure 6.7(a), which shows this matrix around the point in the north-east corner of Figure 6.5 with an altitude of 131 m. Using data from the surrounding points, the slope in the x-direction, from west to east, is found as x-slope = VI/HE = (135 − 115)/200 = 0.1. The value of 200 used is twice

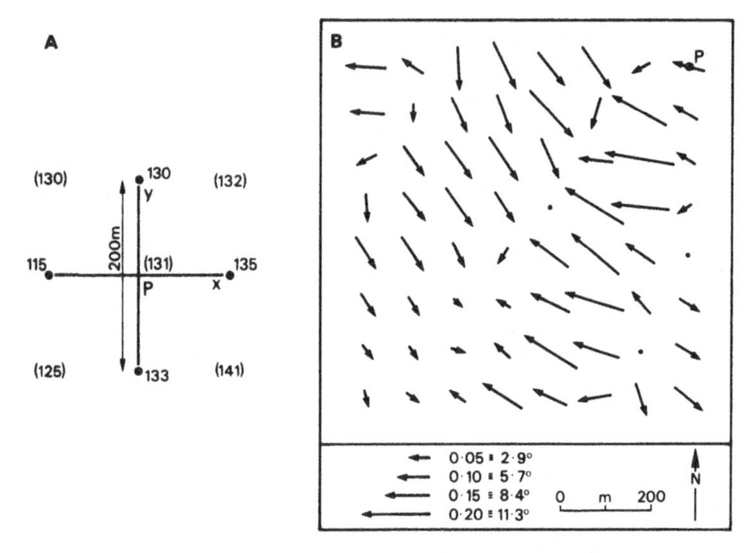

Figure 6.7 a and b Calculation of a gradients map from a scalar field.

the grid spacing. Similarly, the slope in the y-direction (from south to north) is

$$y\text{-slope} = \text{VI/HE} = (130 - 133) / 200 = -0.015$$

The maximum slope, or gradient, at this point is then simply

$$\begin{aligned}\text{gradient at P} &= [(x\text{-slope})^2 + (y\text{-slope})^2]^{0.5} \\ &= [(0.1)^2 + (-0.015)^2]^{0.5} = 0.10112\end{aligned}$$

The direction and sense of this gradient can be found from

$$\begin{aligned}\tan\theta &= x\text{-slope}/y\text{-slope} \\ &= 0.1/-0.015 = -6.667\end{aligned}$$

Finding the angle to which this tangent corresponds is a little tricky, and it may help to refer back to a similar calculation outlined on pp. 77–9 and Table 4.4, reading 'x-slope' for 'sine' and 'y-slope' for 'cosine'. We have a negative tangent and x-slope, together with a positive y-slope, which implies that the required direction lies in the quadrant 270°–360°. From tables, the angular value θ, corresponding to −6.667, is 81.5, so that our gradient has a direction of 360° − 81.5° = 278.5°. While the tangent tables are in front of us, it is worthwhile looking up the slope angle to which our gradient of 0.10112 corresponds. This is 5.7°, so the complete

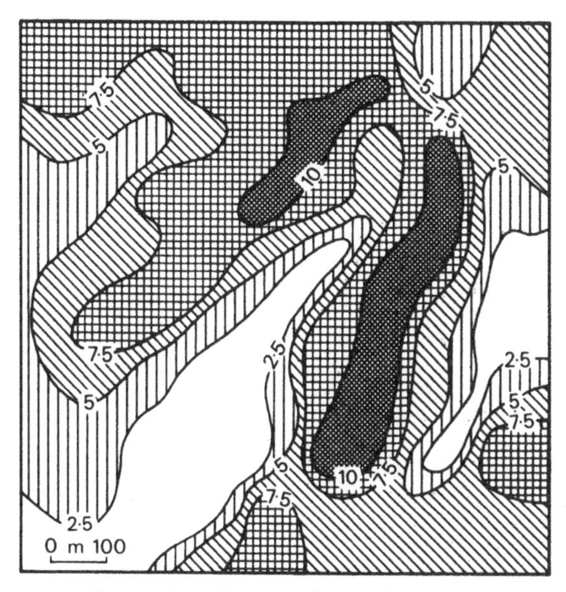

Figure 6.8 The completed slope map.

down-dip gradient is 5.7° at 278.5°. This and the results of a similar calculation for every point in the altitude matrix are shown as proportional arrow symbols in Figure 6.7(b). As might be imagined, the calculations required to get this final map are somewhat laborious, but are easily programmed for a computer. The resulting map should be compared with the original isolines of Figure 6.4; in effect, we have created an objective hachure map. An alternative way of representing the gradients is to draw separate maps for the slope and direction. Figure 6.8 shows the complete slope map, with isolines at 2.5° intervals. Although its derivation does involve rather more labour than is involved in traditional slope mapping, the result is a vastly superior product which summarizes a very important property of the field. We can, of course, now work back through our various measures, using this slope map as the raw material. Of particular interest would be the area/slope relationship and even the slopes of these slopes, that is their straightness or curvature (see Evans, 1972; Dixon, 1971).

INTERPOLATION OF SURFACES

The measures developed in the previous section often assume that the isolines are accurate, both located correctly in plan and with

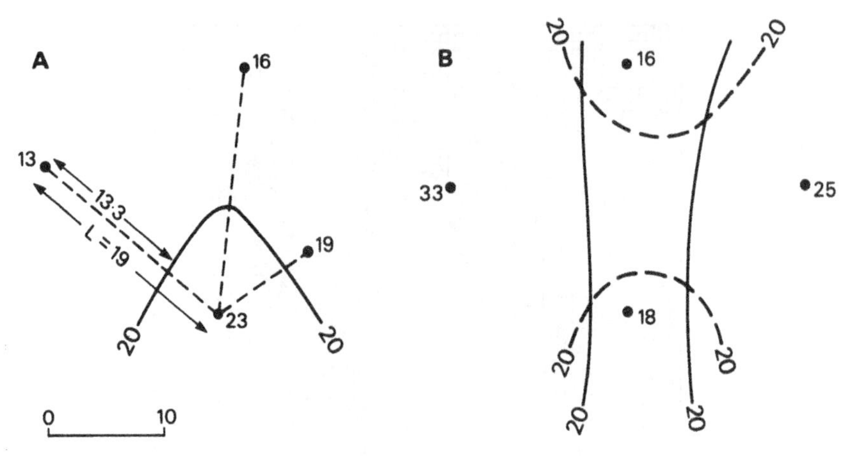

Figure 6.9 a and b Principles of surface interpolation.

precise height values. In virtually all practical problems, this assumption cannot be upheld. Although, in principle, a scalar field such as that of altitude which exists on the ground can be mapped using accurate contours, in practice, cost considerations mean that apparently accurate contours are often actually form lines sketched in between contours in order to give a better visual impression (see Clayton, 1953, for a valuable, but dated, discussion of contour accuracy on Ordnance Survey maps). Even with individually surveyed, accurate contours, there will still be uncertainty about the shape of the ground between them. In studies other than those involving topographic maps, it is simply impossible to contemplate collecting anywhere near as much information. Climatic data, for example, have been collected at a limited number of point locations and, without further survey, are all that we have got to map their spatial variation. It is therefore necessary to draw isolines across the unknown blank spaces between these known control points, a process called *spatial interpolation*.

Traditionally, interpolation of isolines has been done by eye, using a few rules and a rough idea of how the resulting surface ought to look. Figure 6.9(a) illustrates how linear interpolation is used to locate an isoline with a value of 20. From the control point of value 13 to that of value 23 is a distance of 19 units, so that the 20 line can be located an appropriate proportion of the way along, in this case $7/10 \times 19 = 13.3$ units from '13', and so on. Notice that locating isolines in this way involves several assump-

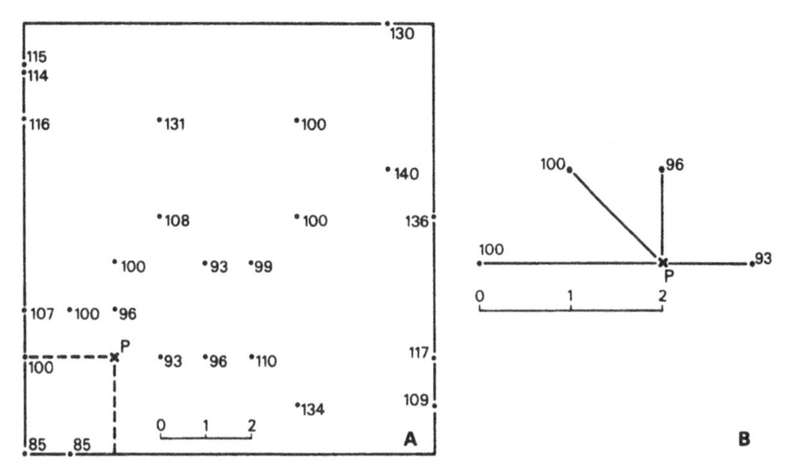

Figure 6.10 a and b A method of objective interpolation.

tions about the surface. It is assumed to be continuous, single-valued and without sharp discontinuities. Moreover, it is possible to isoline only because spatial autocorrelation exists in the data; the closer places are together in distance, the more similar they are in z-value.

At this point the reader should try, in his mind at least, to isoline the distribution of control points shown in Figure 6.10(a). The easiest method is to draw, first, an isoline with a value roughly in the middle of the range of the data, then to work upwards and downwards away from this. In drawing the isolines, it will also be a help to know that the surface mapped is one of altitude, illustrating a very important consideration in isolining, that of the role of prior knowledge. Once it is known that this is relief, we automatically avoid valleys which rise and fall down their long profile and are happy to create sharp V-turns in the contour lines. If we had further been told that the underlying rock is chalk, we may well have tried hard to give a visual impression of rolling downland, and so on.

This 'eyeball' method of interpolation has a number of difficulties. First, as can be established by asking a number of people to isoline the same data, it is not objective. Different people will produce different maps according to their different skills, experience and preconceptions. This problem is particularly acute where the original data are themselves ambiguous, permitting more than one valid isoline pattern. In Figure 6.9(b), which of the two possible isolines of $z = 20$ is correct? Second, because our eyes are trained to seek pattern, there is a danger that we will create patterns of isolines in data which are strictly random. Third, the accuracy of the isolines is controlled by a number of interrelated factors and will not be consistent over the mapped area. The isoline accuracy depends strongly upon the number and distribution of control points, the interval chosen and, more awkwardly, on the unknown characteristics of the surface itself. A further factor is map scale, since on a small-scale map covering a wide area it will be necessary to generalize and smooth the isolines (Pannekoek, 1962).

Fortunately, the advent of the computer has enabled some of these interpolation problems to be overcome and offers the means of producing rapid, objective and consistent interpolation according to any specified criteria or rules. As the review by Rhind (1975) shows, many methods have been developed, ranging from the relatively simple but laborious (at least, by hand) to the extremely complex and even more laborious! To illustrate the principles involved, we shall outline one of the simplest methods – interpolation using inverse distance weighting – which forms the basis of the popular SYMAP computer programme (see Davis, 1973, for a simplified computer programme). The method is outlined in Figure 6.10. Given a pattern of control points, the first step is to lay a very fine mesh of grid points over the area and to interpolate a value at each point on this grid. It will be these interpolated grid values which are eventually isolined. As long as the mesh is fine enough, this interpolation can be done objectively and unambiguously by linear methods. Take the grid point P shown in Figure 6.10(a) at $x = 2$, $y = 2$ as an example. We now find the nearest n points. The precise value chosen for n is a matter of choice, but in the example we select $n = 4$, and these points are indicated in Figure 6.10(b). To find the interpolated value at P, we simply calculate an inverse distance-weighted combination of the z-values of these four neighbours, according to

$$z_p = \sum_{i=1}^{n} (z_i/l_i) \Big/ \sum_{i=1}^{n} (1/l_i)$$

where z = the height of the ith point; l_i = the distance from the point to the grid point P. Table 6.2 completes the calculation for the grid point (2,2), the distances l_i being found using the familiar

Table 6.2 An example of inverse distance weighting (see Figure 6.10)

Neighbour	*Height*	*Co-ordinates*		*Distance*, l_i				*Summation*
i	z_i	x_i	y_i	$(x_i - x_p)$	$(y_i - y_p)$	l_i	z_i/l_i	$1/l_i$
1	100	0	2	−2	0	2	50.0	0.50
2	100	1	3	−1	1	1.414	70.7	0.71
3	96	2	3	0	1	1	96.0	1.00
4	93	3	2	1	0	1	93.0	1.00
							309.7	3.21

Pythagoras method outlined on p. 41. Substitution into the formula gives

$$\begin{aligned} z_p &= (50.0 + 70.7 + 96.0 + 93.0)/(0.50 + 0.71 + 1.00 + 1.00) \\ &= 309.7/3.21 \\ &= 96.5 \end{aligned}$$

Since similar calculations are repeated for every single grid point, and a grid mesh could well have 100 × 100 = 10,000 points, it will be appreciated that the complete interpolation is very much a job for a computer. The final result is illustrated in Figure 6.11, using bands of shading to indicate the contour pattern. It is worth pointing out that the sample of twenty-five control points used were taken randomly from Figure 6.5, so that the completed map ought to look similar to the isoline map in Figure 6.4.

Even with this simple technique, there are at least three ways we could change the rules of the game and so change the final map. One way is to specify a finer or a coarser mesh. A very fine mesh will add a lot of local detail, a coarse mesh will tend to produce a smooth surface. A second way is to alter n, the number of neighbours considered. As n is increased, so points at greater distance are used and we might expect that a smoother map will result. Finally, we can alter the weighting. In the example, use was made of the actual distance l_i ($= l_i^1$). More generally, we could weight using l_i^b, in which the exponent b is set to some value other than unity. Values greater than this will decrease the relative effect of distant points, so producing a bumpier map. Values less than one will increase the importance of distant points. In fact, many computer programmes use $b = 2$, giving an inverse-distance squared weighting.

Given the enormous range of possible objective interpolation schemes, which is the best? The answer is that there is not one

CONTOUR VALUES

SYMBOL	RANGE	
2	85.000-	90.703
	90.703-	96.406
1	96.406-	102.109
	102.109-	107.812
*	107.812-	113.515
	113.515-	119.217
A	119.217-	124.920
	124.920-	130.623
B	130.623-	136.326

Figure 6.11 A computer-generated isoline map.

universally best scheme. Different techniques have different characteristics, and it is up to the analyst to convince himself that the chosen method is best for his particular problem (see Braile, 1978; Morrison, 1974; Rhind, 1971).

TREND SURFACE ANALYSIS

A final property of scalar fields is their *trend*, any large-scale systematic changes that extend smoothly and predictably from one map-edge to the other. Examples of such systematic trends might be the dome of atmospheric pollution over a city, the steady rise in cirque floor levels noted across many mountain ranges, the dome in population density over a city, and so on. The detection and separation of such trends is the object of *trend surface analysis* (see Bassett, 1972; Unwin, 1975).

In the section on pp. 153–4 it was noted that any scalar field can be represented by the equation:

$$z_i = f(x_i, y_i)$$

relating surface height (z_i) to position (x_i, y_i) at each and every data point. As it stands this is pretty vague, since the f simply denotes some function which is unspecified. All trend surfaces do, is to specify a precise mathematical form for this function and fit it to the observed data by least-squares regression. It is unlikely that any simple function will exactly fit the observed data for two reasons. First, even where the real underlying surface is simple, measurement errors will have been introduced into the observed data. Second, in geography it is exceedingly unlikely that only one trend-producing process will be in operation. It follows that as well as a *trend*, represented by a simple mathematical function of the x and y co-ordinates, there will also be local departures from this, or *residuals*. Mathematically, we can denote this as

$$z_i = f(x_i, y_{i},) + e_i$$

i.e. surface height = trend + residual; e_i = residual at the ith data point. Our major problem in any trend surface analysis is to decide upon a particular function for the trend part of the equation. Although there is an enormous range of possible functions, the simplest trend surface we can imagine is an inclined plane, which can be specified as:

$$z_i = a_0 + a_1 x_i + a_2 y_i + e_i$$

Mathematically, the trend is that given by a linear polynomial, and the resulting surface is a *linear trend surface*. To calculate values for the trend part of this equation, all we need to know are the constants a_0, a_1 and a_2 together with the location of the point (x_i, y_i)

and its observed value z_i. The constants have a simple physical interpretation as follows. The first a_0 represents the height of the plane surface at the map origin, where $x_i = y_i = 0$. The second a_1 is the surface slope in the x-direction and the third a_2 gives its slope in the y-direction. Do you recognize a_1 and a_2 as the 'x-slope' and 'y-slope' used on p. 167? As we saw there, different values for these constants will give different gradients for the linear trend surface. All that a least-squares regression does, is to ensure that the constants have values which collectively make the sum of squares of the residuals as small as it possibly can be. It can be shown (Unwin, 1975, pp. 8–9) that these best-fit constants can be found by solving a set of three simultaneous equations, called the normal equations:

$$na_0 + \sum x_i a_1 + \sum y_i a_2 = \sum z_i$$

$$\sum x_i a_0 + \sum x_i^2 a_1 + \sum x_i y_i a_2 = \sum z_i x_i$$

$$\sum y_i a_0 + \sum x_i y_i a_1 + \sum y_i^2 a_2 = \sum z_i y_i$$

where n = total number of data points, $i = 1, \ldots, n$. To solve these, it is necessary to count n, form the sums Σx_i, Σy_i, Σx_i^2, Σy_i^2, $\Sigma x_i y_i$, Σz_i, $\Sigma z_i x_i$ and $\Sigma z_i y_i$, substitute them into the equations and solve for a_0, a_1 and a_2.

The principle involved is fairly straightforward, but in practice the calculations can be exceedingly tedious. To avoid work, and just as important, errors, it is normally advisable to use a computer. Figure 6.12 shows a series of spot heights across a surface to which a linear trend surface is to be fitted. The first step is to measure off the locational co-ordinates (x, y) and to find all the sums required by the normal equations. Table 6.3 shows these calculations. The next step is to substitute these sums into the normal equations, giving:

$$10a_0 + 20a_1 + 20a_2 = 129 \quad (6.1)$$

$$20a_0 + 60a_1 + 39a_2 = 302 \quad (6.2)$$

$$20a_0 + 39a_1 + 64a_2 = 211 \quad (6.3)$$

All that is required to find the one and only combination of values of a_0, a_1 and a_2 which satisfies these equations, is a little school algebra. First, in equation (6.1), express a_0 as

$$a_0 = (129 - 20a_1 - 20a_2) / 10 = 12.9 - 2a_1 - 2a_2$$

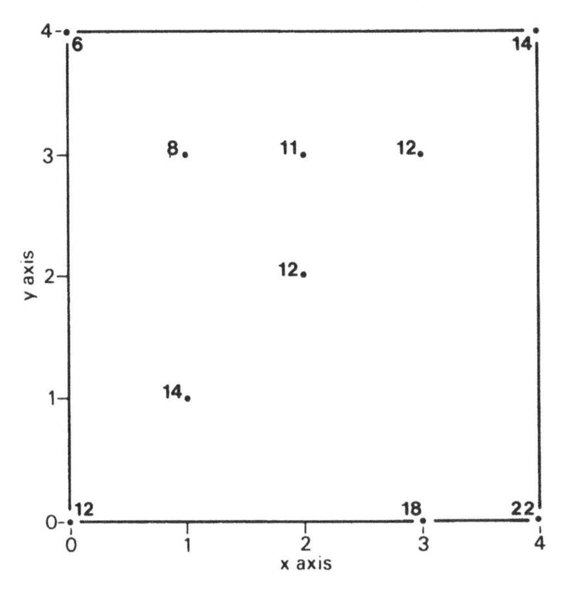

Figure 6.12 Data for a simple trend surface analysis.

then substitute this into both equations (6.2) and (6.3), giving:

$$20\,(12.9 - 2a_1 - 2a_2) + 60a_1 + 39a_2 = 302$$

$$20\,(12.9 - 2a_1 - 2a_2) + 39a_1 + 64a_2 = 211$$

which simplify to

$$20a_1 - a_2 = 44 \quad (6.4)$$

and

$$- a_1 + 24a_2 = -47 \quad (6.5)$$

Continue by finding a_1 from equation (6.4) as:

$$a_1 = (44 + a_2) \,/\, 20 = 2.2 + 0.05a_2$$

and substitute this into equation (6.5) to give:

$$-(2.2 + 0.05a_2) + 24a_2 = -47$$

$$a_2 = -1.8706$$

This is the 'y-slope' of our final least-squares equation. The minus sign indicates that, as y co-ordinates increase, so there is a fall in surface height. Having found one of the constants, it is a simple

Table 6.3 Trend surface analysis summations

i	x_i	y_i	z_i	x_i^2	y_i^2	x_iy_i	x_iz_i	y_iz_i
1	0	0	12	0	0	0	0	0
2	0	4	6	0	16	0	0	24
3	1	3	8	1	9	3	8	24
4	1	1	14	1	1	1	14	14
5	2	2	12	4	4	4	24	24
6	2	3	11	4	9	6	22	33
7	3	3	12	9	9	9	36	36
8	3	0	18	9	0	0	54	0
9	4	4	14	16	16	16	56	56
10	4	0	22	16	0	0	88	0
10	20	20	129	60	64	39	302	211

$\Sigma z_i^2 = 1853$

matter to work backwards, and find:

$$a_1 = 2.1065$$
$$a_0 = 12.4282$$

The complete, best-fitting, linear trend surface is therefore:

$$z_i = 12.4282 + 2.1065x_i - 1.8706y_i + e_i$$

observed height = trend + residual

An isoline map of this surface is shown as Figure 6.13(a), but it is instructive to go a little further and calculate its gradient. As outlined in the previous section, this can be found from the 'x-slope' and 'y-slope', as:

$$\text{gradient} = [(x\text{-slope})^2 + (y\text{-slope})^2]^{0.5}$$
$$= [2.1065^2 + (-1.8706)^2]^{0.5} = 2.817$$

and the direction from:

$$\tan\theta = x\text{-slope}/y\text{-slope}$$
$$= 2.1065/-1.8706 = -1.1261$$

It follows that the angular slope is 70° (tan70° = 2.817) and has a bearing of 311° E of North.

In many studies, it is the form of this trend which is of major interest, but in others, interest may also centre on the distribution of the local residuals. From the previous equations, it is obvious

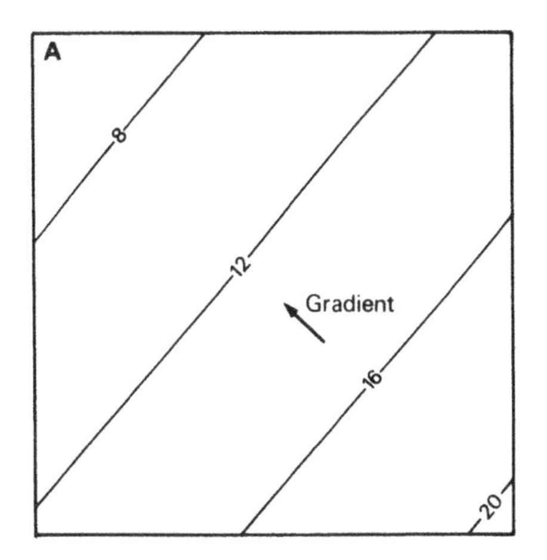

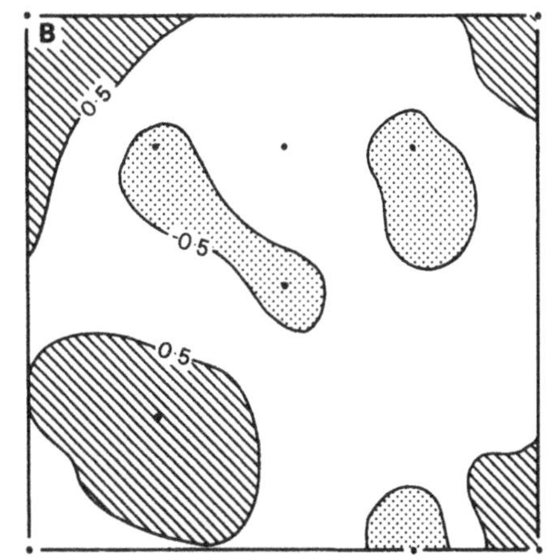

Figure 6.13 a and b Results of a trend surface analysis.

these can be calculated as:

$$
\begin{aligned}
c_i &= z_i - (a_0 + a_1 x_i + a_2 y_i) \\
&= z_i - (12.4282 + 2.1065x_i - 1.8706y_i)
\end{aligned}
$$

Table 6.4 has the necessary calculations. Notice that a check on the computations so far is given by the sum of these residuals, which must equal zero, and the sum of the computed trend values, which must equal the sum of the z_i observed.

Finally, it is customary to derive an index of how well the surface fits the original observed data. This is provided by the *percentage reduction in sums of squares* (%RSS) it achieves, given by:

$$
\%\text{RSS} = \frac{\text{corrected sum of squares of computed trend}}{\text{corrected sum of squares of observed values}} \times 100\ (\%)
$$

$$
= \frac{\sum z_{ti} - \left(\sum z_{ti}\right)^2 / n}{\sum z_i^2 - \left(\sum z_i\right)^2 / n} \times 100
$$

In which z_{ti} denotes the predicted – or computed – value of the trend at the *i*th observation. Substituting the appropriate values from Tables 6.3 and 6.4, gives:

$$
\%\text{RSS} = \frac{(1844.6983 - (129)^2/n)}{(1853 - (129)^2/n)} \times 100 = 95.6(\%)
$$

%RSS values are exactly the same as coefficients of determination in ordinary regression and vary from 100, indicating an exact fit to

Table 6.4 Results of trend surface analysis

Point	x_i	y_i	z_i	*Trend value*	*Residual*
1	0	0	12	12.4282	− 0.4282
2	0	4	6	4.9459	+ 1.0541
3	1	3	8	8.9229	− 0.9229
4	1	1	14	12.6641	+ 1.3359
5	2	2	12	12.9000	− 0.9000
6	2	3	11	11.0294	− 0.0294
7	3	3	12	13.1359	− 1.1359
8	3	0	18	18.7476	− 0.7476
9	4	4	14	13.3718	+ 0.6282
10	4	0	22	20.8541	+ 1.1459
Σ (trend values)				128.9999	0.0000
Σ (trend values)^2				1,844.6983	

0, indicating no fit at all. It is clear that our surface gives a very close fit to the data.

Whether or not this fit is significantly different from zero is tested by calculating the F-ratio

$$F = \frac{\%\text{RSS}/df1}{(100 - \%\text{RSS})/df2}$$

where $df1$ = degrees of freedom associated with fitted surface, equal to the number of constants used, less one for the base term a_0; $df2$ = degrees of freedom associated with the residuals, found as the total degrees of freedom $(n - 1)$, less those already assigned $df1$, that is $n - 1 - df1$. In the example $df1 = 3 - 1 = 2$, and $df2 = 10 - 1 - 2 = 7$, so that

$$F = \frac{95.605/2}{(100 - 95.605)/7} = 76.1403$$

The tabulated critical F-value at the 99.9 per cent confidence level with two and seven degrees of freedom is only 21.69, and since the observed value greatly exceeds this, we can confidently assert that our trend is a real effect and not due to chance sampling from a population surface with no trend.

If this test had revealed no significant trend in the data, several explanations could be adduced. One possibility is that there really is no trend of any sort in the data. Another is that there is a trend in the underlying surface but our sample size n, is too small to detect it. A third possibility is that we have fitted the wrong sort of function. No matter how we change the values of a_0, a_1 and a_2 in our linear trend equation, the result will always be a simple inclined plane. Where this does not provide a significant fit, or where geographical theory leads us to expect a different shape, then other, more complex surfaces may be fitted. Exactly the same techniques are used, but the calculations rapidly become extremely lengthy. Suppose, for example, that we wish to fit a dome or trough-like trend across the study area. The appropriate function to use is a quadratic polynomial, giving a surface:

$$z_i = b_0 + b_1 x_i + b_2 y_i + b_3 x_i^2 + b_4 x_i y_i + b_5 y_i^2 + e_i$$

observed value = trend + residual

There are now six constants, b_0 to b_5, to be found and, therefore, six simultaneous equations. The addition of further terms will give even more complex cubic, quartic, quintic, and so on, trend surfaces but, in practice, these are seldom used (Norcliffe, 1969). Other types of surface, including oscillatory ones, may also be fitted (for reviews see Miesch and Connor, 1968; Davis, 1973; Unwin, 1975).

CONCLUSIONS

If the area-based choropleth map introduced in Chapter 5 is the most popular map type used in human geography, then surface mapping using isolines is the basic tool of the physical geographer. Most data of interest in physical geography, such as altitude, temperature and soil pH are interval or ratio scaled and form truly continuous, single-valued scalar fields. As we have seen, the more general mathematical ideas of scalar and vector fields and matrices can be profitably drawn upon to illustrate this type of mapping. In particular, the use of the surface gradient operator and of other operations of this sort is a powerful but much neglected analytic tool.

Isoline maps are often thought of as being easy to understand and are said to give a good visual picture of the true, underlying surface. Yet, as we have seen, these ideas need to be carefully qualified by an appreciation of their deficiencies. The problems of spatial interpolation and modelling trends still require a great deal of research. In human geography, use is sometimes made of the isopleth map which makes area-based data look as if they came from a true, continuous scalar field. The arguments presented in this chapter imply that, if further quantitative work is to be undertaken using the maps, such representations are inappropriate and should be used only with great care.

RECOMMENDED READING

Cartographic techniques

Keates, J. S. (1961) 'Techniques of relief representation'. *Survey and Mapping* 21, 459–63.

Muehrcke, P. C. (1978) *Map Use: Reading, Analysis, Interpretation,* Madison, JP Publications, especially Chapter 5, 104–26.

Problems and techniques in isopleth mapping are covered in the following series of papers.

Mackay, J. R. (1951) 'Some problems and techniques in isopleth mapping', *Economic Geography* 21, 1–9.

Nordbeck, S. and Rystedt, B. (1970) 'Isarithmic maps and the continuity of reference interval functions', *Geografiska Annaler* B52, 92–123.

Porter, P. W. (1958) 'Putting the isopleth in its place', *Journal of the Minnesota Academy of Science* 26, 372–84.

Schmid, C. F. and MacCannell, E. H. (1955) 'Basic problems, techniques and theory of isopleth mapping', *Journal, Amer. Statis. Assoc.* 50, 220–39.

Tobler, W. R. and Lau, J. (1978) 'Isopleth mapping using histosplines', *Geographical Analysis* 10, 273–9.

The accuracy of isolines is discussed in:

Clayton, K. M. (1953) 'A note on the twenty-five-foot 'contours' shown on the Ordnance Survey 1:25,000 map', *Geography* 38, 77–83.

Pannekoek, A. J. (1962) 'Generalization of coastlines and contours', *International Yearbook of Cartography* 11, 55–75.

General analysis

Bassett, K. (1972) 'Numerical methods for map analysis', *Progress in Geography* 4, 219–54.

Dauphiné, A. (1977) 'Brèves réflexions sur les champs scalaires', *Analyse Spatiale* 3, 35–53.

Davis, J. C. (1973) *Statistics and Data Analysis in Geology,* New York, Wiley.

McQuistan, I. B. (1965) *Scalar and Vector Fields: A Physical Interpretation,* New York, Wiley.

Tobler, W. R. (1967) 'Of maps and matrices', *Journal of Regional Science* 7, 234–52.

Summary measures

Clarke, J. I. (1966) 'Morphometry from maps', in Dury, G. H. (ed.) *Essays in Geomorphology,* London, Heinemann, 235–74.

Clarke, J. I. and Orrell, K. (1958) 'An assessment of some morphometric methods', Durham, University of Durham Occasional Paper No. 2.

Dury, G. H. (1972) *Map Interpretation,* London, Pitman, 4th edn, 167–77.

Evans, I. S. (1972) 'General geomorphometry, derivatives of altitude and descriptive statistics', in Chorley, R. J. (ed.) *Spatial Analysis in Geomorphology*, London, Methuen, 17–90.

Slopes and gradients

Calef, W. and Newcomb, R. (1953) 'An average slope map of Illinois', *Annals, Assoc. Amer. Geogr.* 43, 304–16.
Dixon, R. (1971) 'The direct estimation of derivatives from an irregular pattern of points', *Meteorology Magazine* 100, 328–33.
Raisz, E. and Henry, J. (1937) 'An average slope map of southern New England', *Geographical Review* 27, 467–72.
Thrower, N. J. W. and Cooke, R. U. (1968) 'Scales for determining slope from topographic maps', *Professional Geographer* 20, 181–6.
Tobler, W. R. (1976) 'Spatial interaction patterns', *Journal of Environmental Systems* 6, 271–301.

Interpolation

Braile, L. W. (1978) 'Comparison of four random-to-grid methods', *Computers and Geosciences* 14, 341–9.
Morrison, J. L. (1974) 'Observed statistical trends in various interpolation algorithms useful for first stage interpolation', *Canadian Cartographer* 11, 142–9.
Rhind, D. W. (1971) 'Automated contouring: an empirical evaluation of some differing techniques', *Cartographic Journal* 8, 145–58.
Rhind, D. W. (1975) 'A skeletal overview of spatial interpolation techniques', *Computer Applications* 2, 293–309.

Trend surfaces

Anderson, P. (1970) 'The uses and limitations of trend surface analysis in studies of urban air pollution', *Atmospheric Environment* 4, 129–47.
Bassett, K. A. and Chorley, R. J. (1971) 'An experiment in terrain filtering', *Area* 3 (2), 78–91.
Chorley, R. J. and Haggett, P. (1965) 'Trend surface mapping in geographical research', *Transactions, Inst. Brit. Geogr.* 37, 47–67.
Davies, B. E. and Gamm, S. A. (1969) 'Trend surface analysis applied to soil reaction values from Kent, England', *Geoderma* 3, 223–31.
Lindley, D. V. and Miller, J. C. P. (1962) *Cambridge Elementary Statistical Tables*, Cambridge, CUP, 7 (a) to 7 (c).
Miesch, A. T. and Connor, J. J. (1968) 'Stepwise regression and non-polynomial models in trend analysis', *Computer Contribution* (University of Kansas) 27.
Norcliffe, G. B. (1969) 'On the uses and limitations of trend surface models', *Canadian Geographer* 13, 338–48.

Tarrant, J. R. (1969) 'Some spatial variations in Irish agriculture', *Tijdschrift voor Economische en Sociale Geographie* 60, 228–37.
Unwin, D. J. (1975) 'An introduction to trend surface analysis', Norwich, Geo Abstracts, CATMOG 5.

WORKSHEET

(1) In order to illustrate the idea of a scalar field whose z-value is some function of its (x, y) locational co-ordinates, plot the isoline map given by the function:

$$z_i = 5.8 + 0.12x_i + 2.50y_i + 0.46x_i^2 + 0.09x_iy_i - 0.20y_i^2$$

over the square 'region' from $x = 0$ to 4, and from $y = 0$ to 4.

(2) From the *Daily Weather Report* of the Meteorological Office, abstract a spot-height map of values of temperature over the UK for 0900 GMT on a single day and attempt to isoline these. Having performed this exercise, use random number tables to allocate randomly a series of temperatures in the range 0°–30° and attempt to isoline this random field. Notice how patterns seem to appear!

(3) The standard methods of relief representation used on topographic maps produced by the Ordnance Survey (UK) are well worth detailed examination. First, obtain a 1 : 50,000 map of your own area and abstract a map showing the distribution and values of point height information. How does the density of this information vary? Second, examine the contour patterns used on the following map products – 1 : 50,000, 2nd series, 'Grantown and Cairngorm'; 1 : 63,360, tourist edition, 'Cairngorm'; 1 : 63,360, tourist edition, 'Peak District'; and 1 : 63,360, sheet 111, 'Buxton and Matlock'.

(4) The Ordnance Survey have used the 1 : 63,360 tourist maps as a test-bed for numerous experiments in relief depiction. The following sheets should be examined carefully – 'Wye Valley and Lower Severn', 'Peak Dis-

Continued on next page

trict', 'North York Moors', 'Lake District', 'Loch Lomond and the Trossachs', 'Lorn and Lochaber', 'Cairngorms', 'Dartmoor', 'Exmoor', and 'New Forest'.

(5) Examine the effects of line generalization on the contour patterns shown on 1 : 10,000, 1 : 25,000, 1 : 50,000 and 1 : 250,000 sheets of the area around your home.

(6) A random sample of twenty-five points on the contour map shown in Figure 6.4 yields data of the form

x	y	z
3	5	108
8	6	140
↓	↓	↓
9	5	136

Given that $n = 25$

$$\sum x_i = 94 \qquad \sum x_i^2 = 598$$

$$\sum y_i = 100 \qquad \sum y_i^2 = 564$$

$$\sum x_i y_i = 383 \qquad \sum x_i z_i = 10796$$

$$\sum z_i = 2714 \qquad \sum y_i z_i = 11343$$

substitute these quantities in the normal equations and confirm that the linear trend surface is given by:

$$z_i = 88.4 + 2.3x_i + 2.9y_i$$

Plot the surface over the square from $x = y = 0$ to $x = y = 9$, and observe how it mirrors the evident trend in isolines. Given that the sum of the computed trend values squared $\Sigma z_{ti}^2 = 297410.74$, and the sum of the observed values squared $\Sigma z_i^2 = 300554$, calculate the *%RSS* and assess its statistical significance using the *F*-ratio test.

CHAPTER SEVEN

·MAP COMPARISON·

INTRODUCTION

Although the previous chapters have been concerned solely with the production and analysis of maps to represent distributions of single variables, a great deal of geographical work is concerned with the detection of *common* locational tendencies in two or more distinct variables. Typically, the comparison of maps leads to statements such as 'on the Great Plains, high wheat yields are associated with high rainfall', or more obviously, 'coalmines are located on coalfields'. Both statements assert the existence of an *areal association* between the two variables, both might be used to suggest some linking process, and both can be tested by comparing at least two maps. The number of types of map comparison we can make is enormous. Even if we restrict ourselves to the basic types of distributions, Table 7.1 shows that there are at least ten different types of comparison for which methods might be developed. As the table shows, comparisons between similar types of distribution, numbers 1, 5, 8 and 10, are probably the most obvious. Examples are comparisons of the point patterns of central places with those predicted by a Christaller model (type 1), the line patterns of the road and railway networks of an area (type 5), or the area distributions of woodland and a particular soil type (type 8). Nevertheless, comparisons of distributions of differing spatial dimension are often undertaken, as for example a type 3 comparison of a point pattern of farms with the soil type's area pattern or the distribution of wooded areas with a surface of altitude (type 9), and so on. This picture is further complicated when we consider the measurement level of the data: within each of the ten types the data used can be either nominal, ordinal, interval or ratio.

Given the importance of map comparison to geographical enquiry, it might be expected that the methodology of comparison

Table 7.1 Possible types of map comparison

Distribution B	*Distribution A*			
	Point	*Line*	*Area*	*Surface*
Point	1	2	3	4
Line		5	6	7
Area			8	9
Surface				10

Note: Within each category of distribution, data can be nominal, ordinal, interval or ratio.

would be well developed and understood, yet this is not the case; the literature on map comparison is very sparse indeed (see Robinson and Bryson, 1957; Tobler, 1965; Court, 1970; and Cliff, 1970, for exceptions).

VISUAL COMPARISON

The simplest way to compare two maps is, of course, to look at them, or to overlay them and describe how their distributions vary together. Visual map comparison has been a basic method for many years and is often very effective. Figures 7.1(a) and (b) show the ward-by-ward distributions of overcrowding (defined as a density greater than 1.5 per room) and the percentage of homes rented privately over the city of Leicester in 1971. It can be seen that there is a strong areal association between the two which might lead us to hypothesize that they are in some way causally related, but this association is by no means exact.

A fundamental problem in this method lies in its subjectivity. Confronted with the same two maps, not everyone will agree that an areal association exists, or will assess the degree of association as the same. If the association is strong and positive, most people will reach a similar conclusion, but with weak, or even negative, associations this is no longer true. Visual map comparison is further complicated by the maps themselves. It is necessary to ensure that like is compared with like, and the extent of association inferred can be related as much to the techniques of mapping used as to the realities of the underlying geography. In both choropleth and isoline mapping, for example, it is vital to ensure that the class interval and number used are the same (see Muller, 1967; Olson, 1972; Lloyd and Steinke, 1977).

Although most visual map comparison uses overlaid maps, it is

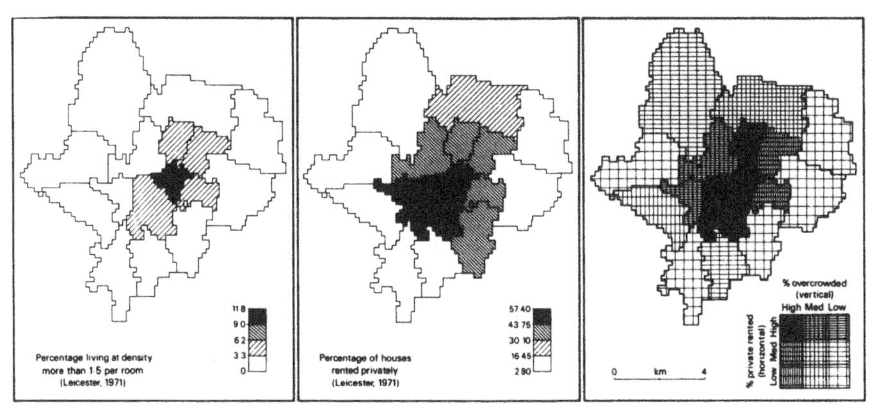

Figure 7.1 a, b and c Overcrowding and private rented houses, Leicester, 1971.

possible to use symbolism to indicate the areal correspondence of two distributions, using a development of the method pioneered by Learmonth and Pal (1959) which is applicable to comparisons involving area or surface data, types 8, 9 and 10. This is illustrated in Figure 7.1(c), which uses the same Leicester data but has a two-way categorization of class intervals into tertiles as shown in the key. One variable is shown using vertical shading, the other horizontal, so that the relative darkness of the resulting shades gives a crude visual assessment of their areal association. The inner-city wards of Wycliffe, Latimer and Castle stand out as being high on both variables, while the suburbs of Aylestone, Newton, Humberstone and Evington show equally low values on both variables and the remaining wards show intermediate associations.

NUMERICAL COMPARISON

The difficulty with an approach, no matter how sophisticated, which uses visual comparison is that the result is subjective and there is no way of assessing the strength of association. To overcome this, a number of measures have been proposed and illustrated but, as yet, these do not seem to have been extensively used in research.

The coefficient of areal correspondence

The simplest numerical coefficient of use in map comparison is that proposed by Minnick, R. F. (1964) to evaluate the correspon-

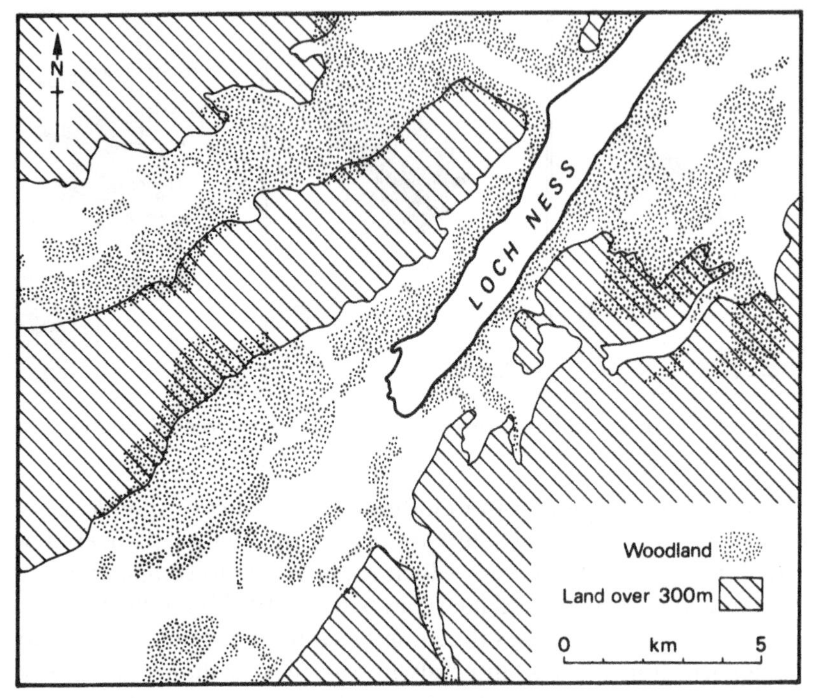

Figure 7.2 Woodland and land above 300 m around Fort Augustus, Scotland. (Source: Bartholomew, 1:100,000, Sheet 51.)

dence between natural area patterns (type 8), between areas and surfaces (type 9), or with modification, between surfaces (type 10). Suppose that we have overlain two maps of area data as shown in Figure 7.2, which depicts the distributions of woodland and high and low land around Fort Augustus at Loch Ness, Scotland. Visually, it is apparent that woodland tends to be restricted to the lower land, below 300 m, but the *coefficient of areal correspondence* quantifies this association as:

$$C_a = \frac{\text{area over which phenomena are located together}}{\text{total area covered by the two phenomena}}$$

C_a varies from 0 to 1. Completely separate distributions give a value of zero, while exactly coincident ones give a value of unity. The areas can be found by any suitable method and organized into a table (Table 7.2). The numerator in the expression for C_a is simply the wooded area below 300 m, 68.3 km^2, and the denominator is

Table 7.2 Quantities needed to calculate C_a, the coefficient of areal correspondence, woodland and lowland around Fort Augustus (km^2)

	Altitude		
	less than 300 m	*more than 300 m*	*Total*
Wooded	68.3	14.3	82.6
Unwooded	72.3	150.2	222.5
	140.6	164.5	305.1

Note: Loch Ness, 14.9 km^2 in area, has been omitted from the calculation.

this area plus both the unwooded land below 300 m (72.3 km^2) and the wooded land above 300 m (14.3 km^2), giving:

$$C_a = 68.3 / (68.3 + 72.3 + 14.3) = 0.441$$

and indicating a reasonable degree of areal correspondence.

A modification and development of this method which allows it to be used for surface comparisons of type 10 has been proposed by Court (1970). His method is to draw the same number of isolines at the same, standardized, intervals on each map, then to overlay the two. Clearly, the isolines will usually intersect to give several discrete regions. The measured areas of these regions are then gathered into a *resemblance matrix* from which a coefficient of areal correlation is calculated. In the simplest case only one isoline is drawn at the median value of each distribution, giving two maps of high and low values. Superimposition of these maps gives a map with four types of region in which both variables are respectively: above the median (++), below (−−); one above and the other below (+−); and one below, the other above (−+). The coefficient of areal correlation is given by:

$$C_m = \frac{\text{sum of like areas} - \text{sum of unlike areas}}{\text{total area of region}}$$

Court (1970) gives the example of a comparison of two isopleth maps of south-east Oregon showing the percentage of families with low income and the percentage employed in manufacturing. The resemblance matrix, in percentage units, is shown in Table 7.3 and the coefficient C_m has the value

$$\begin{aligned} C_m &= [(19.4 + 16.2) - (35.9 + 285)]/100 \\ &= -0.288 \end{aligned}$$

Table 7.3 Resemblance matrix for two isoline maps

		Percentage low income	
		above (+)	*below*(−)
Percentage in manufacturing	*above* (+)	19.4	28.5
	below (−)	35.9	16.2

Source: Court (1970), 436.

This coefficient has perhaps three advantages over C_a. First, its limits are −1 to +1 with a perfect negative areal correspondence, giving a value of −1. Second, as long as the number of classes used is even, the method can be extended to deal with maps with 4, 6, . . . , *k* classes. Finally, according to Court, the sampling distribution of his coefficient is roughly normal, so that if required, it is possible to test for statistical significance (see for details, Court, 1970, p. 436.)

The Lorenz curve

Because they are based on the correspondence of areas, both the Minnick and Court coefficients should not be used for area comparisons of type 8 (Table 7.1), when the areas used are imposed ones such as counties as in a comparison of choropleth maps. When the areas used are imposed, it makes little sense to use them in finding an index of map correspondence. For choropleth maps a more useful approach is to construct a Lorenz curve, which is an area-by-area plot of the ratios of the two variables made in order to indicate similarities of distribution. The calculations involved can be illustrated using data for Sheffield in 1971 (Figure 5.10), relating to the number of owner-occupied homes (z_1) and the number of households in each ward with two or more cars (z_2). These data are shown in Table 7.4 and Figure 7.3 which confirms what we might expect, that wards with a high number of owner-occupied homes will also have a high number of multicar households; both are indicative of areas of high social status. To calculate a Lorenz curve, we proceed as follows. First, the ratio z_1/z_2 of the two variables is calculated for each area, and, second, these ratios are ranked with the lowest given rank one. Next, each variable is standardized by expressing it as a percentage of its

Table 7.4 The Lorenz curve: numbers of owner-occupied houses (z_1) and households owning two or more cars (z_2), Sheffield, 1971

Ward	z_1	z_2	z_1/z_2	*Rank*	$\%z_1$	$\%z_2$	*Cumulative*	
1 Attercliffe	1,877	87	21.6	27	2.6	0.9	100.0	100.0
2 Beauchief	5,739	988	5.8	6	7.9	10.1	26.5	42.9
3 Birley	2,582	351	7.4	13	3.6	3.6	51.0	70.9
4 Brightside	2,735	181	15.1	23	3.8	1.9	82.5	92.6
5 Broomhill	3,312	763	4.3	4	4.6	7.8	11.8	23.5
6 Burngrave	1,970	162	12.2	19	2.7	1.7	67.4	84.4
7 Castle	771	105	7.3	12	1.1	1.1	47.4	67.3
8 Darnall	3,832	208	18.4	26	5.3	2.1	97.4	99.1
9 Dore	3,958	1,203	3.3	1	5.5	12.4	5.5	12.4
10 Ecclesall	5.785	979	5.9	8	8.0	10.1	38.0	57.4
11 Firth Park	2,054	156	13.2	21	2.8	1.6	76.1	89.4
12 Gleadless	2,516	431	5.8	7	3.5	4.4	30.0	47.3
13 Hallam	4,899	909	5.4	5	6.8	9.3	18.6	32.8
14 Handsworth	2,491	321	7.8	14	3.4	3.3	54.4	74.2
15 Heeley	3,740	231	16.2	24	5.2	2.4	87.7	95.0
16 Hillsborough	4,281	333	12.9	20	5.9	3.4	73.3	87.8
17 Intake	3,272	368	8.9	17	4.5	3.8	63.2	81.7
18 Manor	558	148	3.8	2	0.8	1.5	6.3	13.9
19 Mosborough	1,235	191	6.5	9	1.7	2.0	39.7	59.4
20 Nether Edge	3,932	538	7.3	11	5.4	5.5	46.3	66.2
21 Nether Shire	676	171	4.0	3	0.9	1.8	7.2	15.7
22 Netherthorpe	1,106	100	11.1	18	1.5	1.0	64.7	82.7
23 Owlerton	1,812	207	8.8	16	2.5	2.1	58.7	77.9
24 Park	844	127	6.7	10	1.2	1.3	40.9	60.7
25 Sharrow	1,897	128	14.8	22	2.6	1.3	78.7	90.7
26 Southley Green	1,290	152	8.5	15	1.8	1.6	56.2	75.8
27 Walkley	3,184	196	16.2	25	4.4	2.0	92.1	97.0
	72,348	9,734			100.0	100.0		

overall total and, *maintaining the ranks*, these percentage values accumulated in the manner shown in the final two columns of the table. Finally, these cumulative percentages are plotted against each other, as shown in Figure 7.4. How does this curve help in map comparison?

If you re-examine each step in the procedure, it can be seen that the resulting plot will be a straight, diagonal line if, and only if, the distributions are proportionately identical in each area. Complete separation of the distributions will give a line which follows the z_2 axis exactly and any real-world distributions of interest, like the

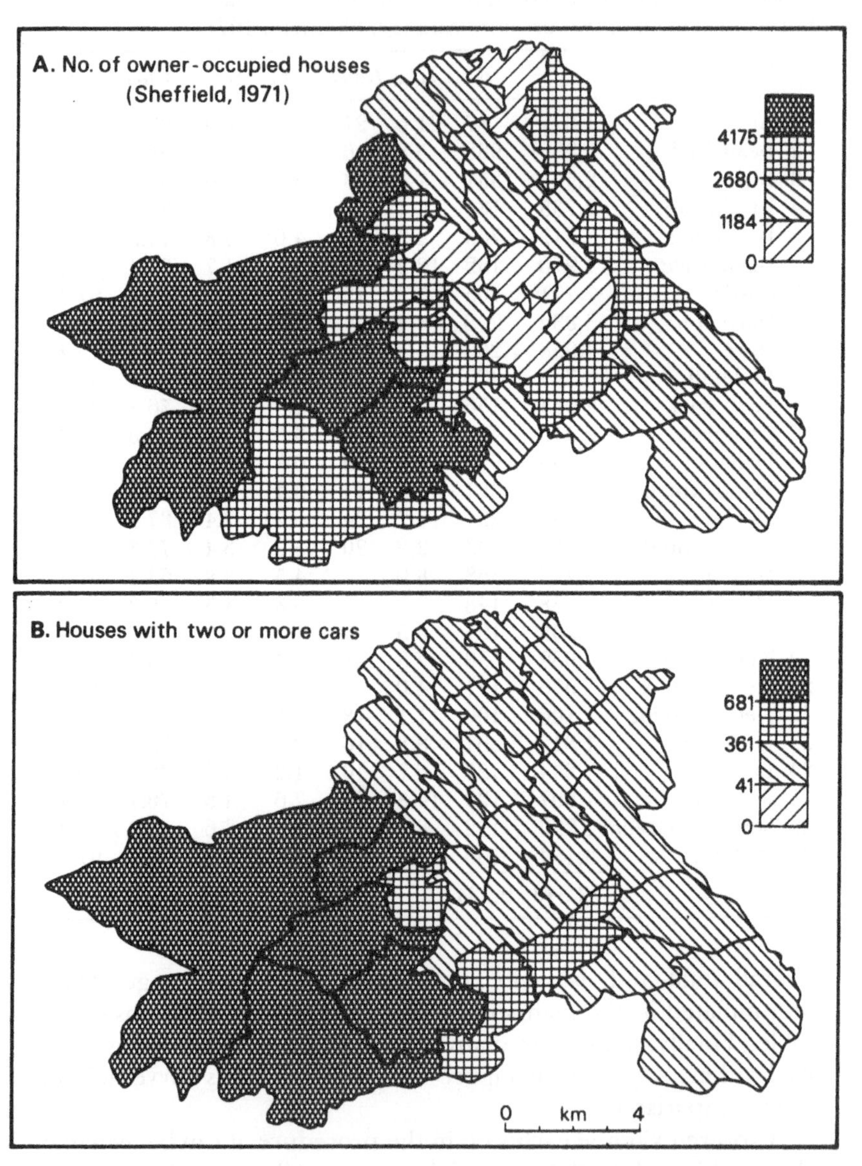

Figure 7.3 a and b The distribution of owner-occupied houses and households owning two or more cars, Sheffield, 1971.

Sheffield data, will lie somewhere between these extremes. It can also be seen that the difference between the actual plotted curve and the theoretical ideal is a measure of the dissimilarity of the distributions.

In fact, the *index of dissimilarity* is usually found as the maximum vertical distance separating the two curves, expressed as a percentage of the maximum possible. This can be measured directly or found from the data as the maximum value of the differences in cumulated percentages. In our example, it occurs at Birley ward and has the value:

$$D_I = 70.9 - 51.0 = 19.9$$

A second method of finding D_I is from the sum of the absolute differences (that is, ignoring the signs) of the individual percentage values divided by two;

$$D_I = \sum |\%z_1 - \%z_2| / 2$$
$$= 40.0/2 = 20.0(\%)$$

A third method finds an index of dissimilarity from the ratio of two areas. In Figure 7.4 the area between the curves is a measure of how poor the fit is, while the total area below the diagonal is that which would obtain for the worst possible fit. A simple index is, therefore, the ratio of these areas:

$$D_I = 1352 \ / \ 5000 = 0.27.$$

This value is numerically higher than the previous two, but it makes better use of the available information; in all three cases the low values of dissimilarity indicate that the distributions correspond reasonably well. The index of dissimilarity, or one very like it, appears in the literature in a variety of guises and has been given different names according to the nature of z_1 and z_2. In population studies, with z_1 as the population and z_2 the land area, it is the *index of population concentration*. With the same z_2, but with z_1 the land area under a particular use, it becomes the even better-known *coefficient of areal localization*, and so on.

The Lorenz curve is a useful graphical and numerical way of making map comparisons, but it suffers from a number of disadvantages. It can only be used for choropleth maps and is open to all the problems of areal unit size and number that were outlined in Chapter 5. It should not be used with negative numbers, or with

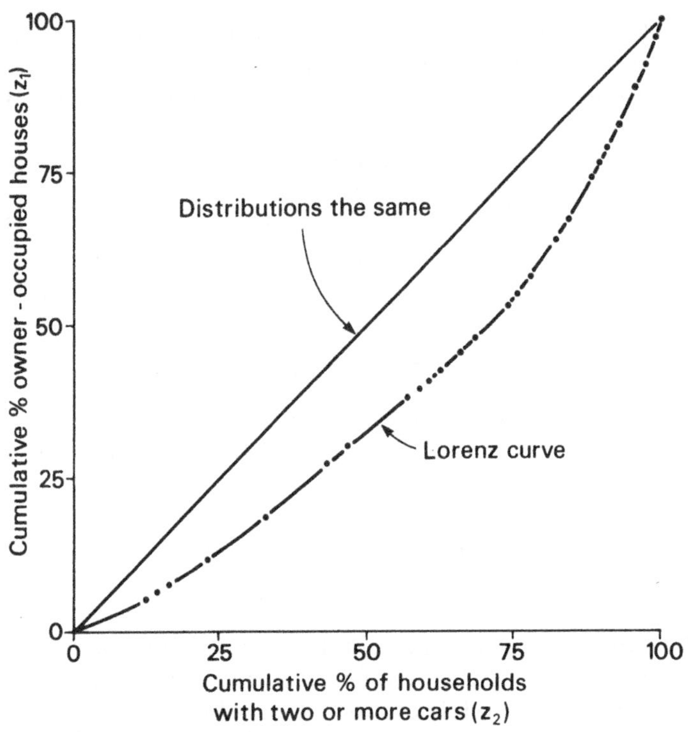

Figure 7.4 Lorenz curve for the Sheffield data.

numbers which are themselves density ratios, and it will not discriminate between different arrangements of the areal units involved.

The correlation coefficient as areal association

A more general approach which gets over some of these difficulties is to use the statistical correlation coefficient as a measure of areal association, in the manner suggested by Robinson and Bryson (1957). The details are outlined in virtually every standard statistical text but are worth a brief discussion in order to show how the statistical idea is translated into a map comparison. The critical quantity calculated in the product moment correlation coefficient r is the covariance, which is a measure of how the variables vary together away from their respective means. It is found from:

$$\text{covariance}(z_1 z_2) = \sum_{i=1}^{n} (z_{1i} - \bar{z}_1)(z_{2i} - \bar{z}_2)/n$$

where n = number of observations; $\bar{z}_1, \bar{z}_2$ = respective means. It can be seen that the covariance will be high and positive when *both* z_1 and z_2 values are above or below their respective means, and it will be a large negative value when one value is below its mean, the other above. In both cases the result, if most of the z_1 and z_2 vary together in the same way, will be to give a large (positive or negative) covariance. When systematic covariation is absent, the covariances will tend to cancel out, giving a final sum close to zero. As it stands, this covariance depends strongly on the units used to measure z_1 and z_2, and in order to produce an index varying from -1 to $+1$ it is usual to standardize it by dividing by the product of the two standard deviations s_{z_1} and s_{z_2}, each calculated as :

$$s_{z_1} = \left[\sum (z_{1i} - \bar{z}_1)^2 / n\right]^{0.5}$$

$$s_{z_2} = \left[\sum (z_{2i} - \bar{z}_2)^2 / n\right]^{0.5}$$

This yields the product moment correlation coefficient r as:

$$r = \text{covariance}\ (z_1 z_2)\ /\ s_{z_1} s_{z_2}$$

The statistical correlation can be seen as rather similar to the geographical coefficient of areal correspondence C_a, but with the *height* or z-value in place of the area ones used by C_a. Like C_a, it is a ratio of the way two variables vary together divided by the total possible variation in both. Because it uses height, rather than area, the r-values obtained do not depend upon the spatial units over which the data were collected but, if desired, they can be made to do so, by weighting each pair of values according to the proportion of the total area they represent.

For the Sheffield data dealt with in the previous section, the product moment correlation coefficient can be found from:

$$n = 27$$

$$\bar{z}_1 = 2679.6 \qquad \bar{z}_2 = 360.5$$

$$\sum (z_{1i} - \bar{z}_1)^2 = 60391699 \qquad \sum (z_{2i} - \bar{z}_2)^2 = 2764160$$

$$s_{z_1} = [60391699/27]^{0.5} = 1496 \qquad s_{z_2} = [2764160/27]^{0.5} = 320$$

$$\text{covariance}\ (z_1 z_2) = 371449$$

$$r = 371449/1496 \times 320 = 0.776$$

The close relationship between the two maps, evident by visual comparison, is confirmed. This correlation coefficient makes use of interval or ratio scaled data, but it can equally well be replaced by other suitable coefficients, such as Spearman's rank coefficient for ranked data or even one or other of the various coefficients used to measure association in nominal data described in Siegel (1956). As we have described it, it is further restricted to comparisons of scalar quantities of the type found on proportional symbol or choropleth maps but it can be extended to enable vectors to be correlated, thus permitting the direct comparison of pure point (type 1) and line patterns (type 5). An introduction to vector correlation is given by Tobler (1965), to which the interested reader is referred.

Because it makes use of standard and well-understood procedures in its computations, the correlation approach to areal association is both simple and attractive. There are, however, two major traps for the unwary in its use. The first is the possibility of confusing covariation with causation and inferring that a high correlation coefficient necessarily means that the two maps are causally related. True, they may be, but the correlation does not prove it; all we can say is that similar relative values tend to occur at similar locations. Second, having calculated a coefficient, the temptation is to test it for statistical significance, that is to compare the observed r-value with the theoretical distribution of values expected by chance from a population whose underlying correlation is zero. This is not permissible. Statistical significance testing makes a number of assumptions, notably that of independent random sampling, that cannot be sustained in any map comparison work. The z-values used are in most cases the total population available, so that sampling as a statistician understands it is not involved. Even if we argue, as on p. 119, that the act of imposing data-collection zones is a form of sampling, this is almost never random and, as yet, statistician-geographers have only limited understanding of its effects. Finally, it is likely that the distributions compared will both show significant spatial autocorrelation, thus in no sense is our sample an independent one, and it can be shown that when positive autocorrelation is present, it leads to overstatement of the statistical significance of the results. By all means, use correlation coefficients as measures of the similarity of mapped patterns, but do not invest them with a further inferential significance.

Despite these problems, the correlation approach has the merit

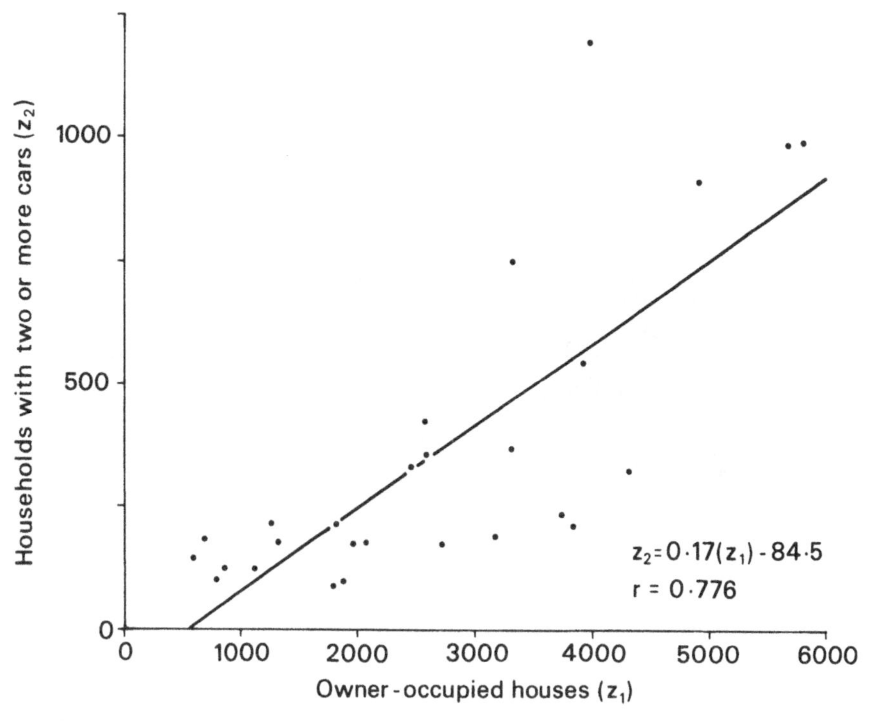

Figure 7.5 Regression of number of households owning two or more cars (z_2) on number of owner-occupied houses (z_1), Sheffield, 1971.

of allowing the response of one dependent variable to changes in the value of the other to be calculated using least-squares regression. Maps of the residuals from the regression provide an often-used and valuable diagnostic tool in isolating the effects of other variables (Thomas, 1968). The procedure is to calculate, first, a least-squares regression line of best fit, then predict values of the dependent variable from this. For the Sheffield data the most sensible way of doing this is to predict the number of households with two or more cars (z_2) from a knowledge of the number of owner-occupied houses (z_1). Using the method outlined in Table 4.6, the regression is

$$\text{Predicted } z_2 = 0.17(z_1) - 84.5$$

and this has been plotted in Figure 7.5. As expected, it shows a steady rise in multicar families as the number of owner-occupied houses increases in each census ward. Notice again that the

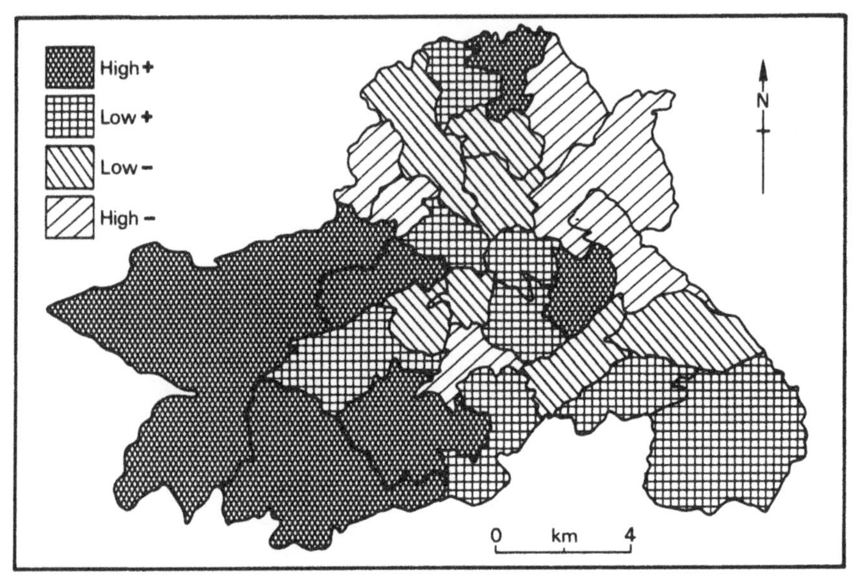

Figure 7.6 Residuals over regression of number of households owning two or more cars on number of owner-occupied houses, Sheffield, 1971.

relationship is not directly causal; both variables reflect a common underlying dimension of prosperity.

Although the fitted line is in itself of interest, from a geographical point of view, it is even more interesting to examine the positive and negative departures, or residuals, of the observed data from it. For example, Attercliffe had 1877 owner-occupied houses in 1971, and the regression predicts that it should, therefore, have:

$$\text{Predicted } z_2 = 0.17(1877) - 84.5 = 234.6$$

households with two or more cars. (Notice in passing that our data are integer values, or whole numbers ranging from zero upwards. The observation that we cannot have 0.6 of a household merely serves to remind us of another assumption we have broken, that values of the variable should be continuous and unbounded.) In fact it had only 87, giving a residual of −147.6.

Figure 7.6 shows the spatial distribution of all the residual values: wards which have more multicar families than expected are in the outer western, high-status suburbs notably Beauchief, Dore and Hallam together with inner-city wards of Nether Shire and Manor which have a large proportion of council-owned housing. Wards with fewer multicar households than expected are in the depressed inner city, notably Brightside, Attercliffe and Darnall in the east.

THE DIFFERENCE MAP APPROACH

The various map comparison methods outlined in the previous section are alike in several respects. In every case they measure the similarity of two patterns by a single number, but do not explicitly consider the spatial arrangement of this similarity. It is perfectly possible, for example, to have a high correlation between two maps even when the spatial arrangement of the differences between these maps is anything but random. As we have seen, none of the proposed measures has successfully been linked with statistical theory to give a statistical test for the similarity of two maps. A method which answers these difficulties has been pioneered by Cliff (1970) and is based on the analysis of the *pattern of differences between two maps*. Suppose that we have two maps based on the same areal units. One might be an observed map, the other derived theoretically, or both might be observed, real-world patterns. All that matters is that they are measured in the same units and relate to the same areas. The important step in deriving a test is to construct a map of the *differences* between the two. Most simply, this is accomplished by creating a three-colour map on which each areal unit is coloured according to the relationship between the two values. By convention, the maps are shaded black (B) if the two values are the same, white (W) if the value on the first map is greater than on the second and red (R) if it is less. If the original data are interval or ratio scaled, exact equalities will be rare, and attention concentrated on the distributions of W and R joins, but with ordinal data such as choropleth categories, many areal units may well be coloured B. The fundamental proposition in Cliff's method is that if the two maps do not differ significantly in a spatial sense, then the distributions of the three colours will not be significantly different from random. On the other hand, any spatial pattern in these colours will indicate the operation of some other unknown spatial process. All that is necessary is a test for randomness in a three-colour map, and it turns out that a simple joins-count approach, similar to that outlined on pp. 135–45 can be applied. As in the normal joins-count test for spatial autocorrelation, the test statistic is a standard normal deviate, or z-score, defined as:

$$z = \frac{\text{observed number of joins} - \text{expected number given by a random pattern}}{\text{standard deviation of expected values}}$$

The observed values can be obtained by counting and, for joins of the same colour (BB, RR and WW), the expected number for a particular colour is:

$$J_c = kn_c^{(2)}/n^{(2)}$$

where J_c = expected number of joins of the same colour c; n_c = number of joins of same colour c; n = total number of joins; $n^{(x)} = n(n-1)\ldots(n-x+1)$. The standard deviation of this number is given by the expression

$$[J_c + 2mn_c^{(3)}/n^{(3)} + \{k(k-1) - 2m\}\, n_c^{(4)}/n^{(4)} + J_c^2]^{0.5}$$

in which k and m are defined as on p. 137:

$$k = 0.5\sum_{i=1}^{n} j_i$$

$$m = 0.5\sum_{i=1}^{n} j_i(j_i - 1)$$

where j_i = number of joins to the ith county. These expressions will be recognized as modifications of the non-free sampling joins-count test for spatial autocorrelation detailed on pp. 140–1. Expressions for the expected number and standard deviation of joins of differing colour (BW, BR and RW) are more complex and can be found in Cliff's (1970) work.

Figure 7.7 is a two-colour map of the differences between a map of the ward proportion of housing at density greater than 1.5 per room and a map of the proportion of privately rented housing over the wards of Leicester in 1971. Areas shaded 'red' are those over which there is proportionately less overcrowding, areas left 'white' have proportionately more. If the original two maps differ significantly, we would expect the pattern of shaded (R) and unshaded (W) areas to be non-random. The required counts are

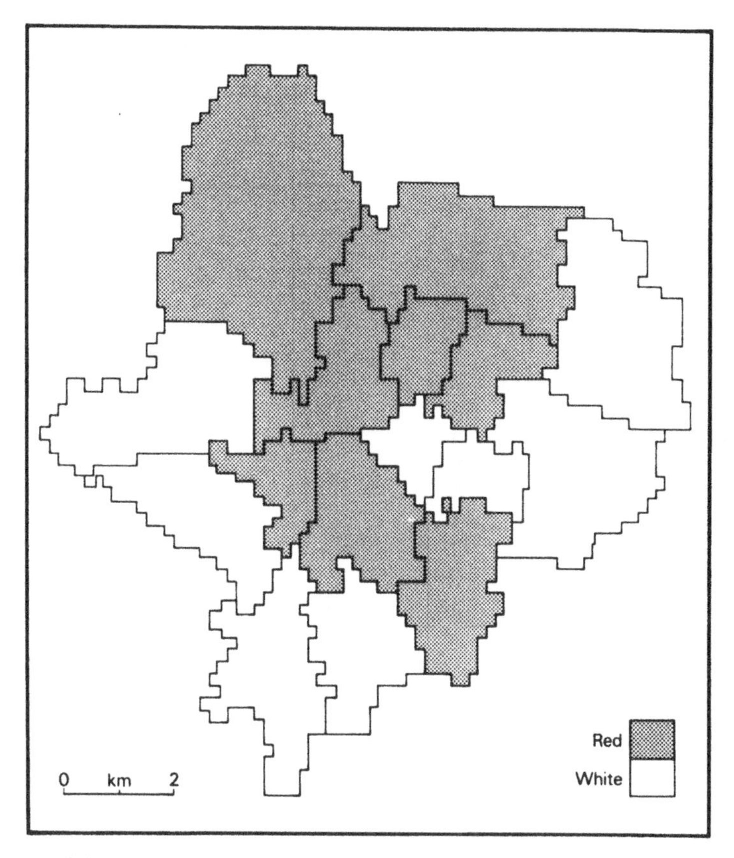

Figure 7.7 Differences map derived from Figure 7.1, Leicester, 1971.

shown in Table 7.5, from which k, m, J_{WW} and the standard deviation of J_{WW} can be calculated as:

$$k = 0.5(72) = 36$$

$$m = 0.5(278) = 139$$

Taking joins of type WW as an example, we have:

$$J_{WW} = kn_w^{(2)} / n^{(2)} = 36\ (8)\ (7) / (16)\ (15) = 8.4$$

standard deviation of J_{WW}

$$= [J_{WW} + 2mn_w^{(3)}/n^{(3)} + \{k(k-1) - 2m\}n_w^{(4)}/n^{(4)} - J_{WW}^2]^{0.5}$$

$$= [8.4 + 278(336)/3360 + \{36(35) - 278\}1680/43680 - 8.4^2]^{0.5}$$
$$= 3.409^{0.5} = 1.846$$

Table 7.5 Cliff's (1970) joins-count approach to map comparison, Leicester, 1971

			Number of joins of		
Ward	*Colour*	*j*	RW	WW	RR
The Abbey	R	3	1	0	2
Belgrave	R	5	1	0	4
Newton	W	4	3	1	0
North Braunstone	W	3	1	2	0
St Margarets	R	7	2	0	5
Latimer	R	4	1	0	3
Charnwood	R	6	4	0	2
Wycliffe	W	5	4	1	0
Spinney Hill	W	5	3	2	0
Knighton	R	4	3	0	1
Humberstone	W	3	2	1	0
Evington	W	4	2	2	0
Castle	R	7	4	0	3
Westcotes	R	5	3	0	2
Aylestone	W	4	2	2	0
De Montfort	W	3	2	1	0
		72	38	12	22
n_r =	8	36	19	6	11
n_w =	8				

Hence, the standard normal deviate used to test for significance

$$z_{WW} = (\text{observed number of WW joins} - J_{WW}/sd(J_{WW})$$
$$= (6 - 8.4)/1.846$$
$$= -1.300$$

which is not statistically significantly different from random at the 95 per cent level. The expected value and standard deviation for RR joins are the same as those for WW ($n_r = n_w$), but the observed value is 11, giving:

$$z_{RR} = (11 - 8.4) / 1.846 = 1.408$$

which is also not significantly different from random. We conclude that we have no evidence in the difference map to assert that the original two maps significantly differ spatially.

The test reported above is both useful and important, enabling a statistically sound approach to be made to the fundamental problem of map comparison but, like the joins-count test on which it is based, it loses a lot of information which was present in the original data. There is, however, no reason at all why the same approach should not be used with a different, more powerful test for spatial autocorrelation and the idea of maps of difference between distributions is one which is of very general utility.

CONCLUSIONS

As we have seen, map comparison lies at the heart of the geographical method in which the areal association of phenomena is used to suggest possible causal links between them. Usually and most obviously, such comparison will involve maps of different variables, for example, rainfall and wheat yield, but in a sense virtually all the tests reported in the previous chapters also involve map comparison. The general procedure has been to compare measures made on observed, real-world patterns of a single variable with the results of a postulated process, usually the independent random. On occasions, maps of a single realization of such a process have been displayed but, more often than not, these theoretical 'maps' remained implicit in the mathematics of the calculation. Even so, map comparison still is involved.

In considering the results of any application of this geographical method of map comparison, we can do no better than to return to Chapter 1 (p. 6) on what maps cannot do, and remind ourselves that the demonstration of a similarity between two maps does not establish a causal link. The same forms can be generated by markedly different processes and only seldom will any simple analysis be able to discriminate between them. This observation should not be taken to the conclusion that the drawing and analysis of maps is scientifically worthless, on the contrary, it forms an important step on the way towards any geographical explanation.

RECOMMENDED READING

Bachi, R. (1963) 'Standard distance measures and related methods for spatial analysis', *Papers, Regional Science Association* 10, 83–132.

Cliff, A. D. (1970) 'Computing the spatial correspondence between geographical patterns', *Transactions, Inst. Brit. Geogr.* 50, 143–54.

Court, A. (1970) 'Map comparisons', *Economic Geography* 46, 435–8.

Learmonth, A. T. A. and Pal, M. N. (1959) 'A method of plotting two variables (such as mean incidence and variability from year to year) on the same map, using isopleths', *Erdkunde* 13, 145–50.

Lewis, P. (1977) *Maps and Statistics*, London, Methuen.

Lloyd, R. and Steinke, T. (1977) 'Visual and statistical comparison of choropleth maps', *Annals, Assoc. Amer. Geogr.* 67, 429–36.

Minnick, R. F. (1964) 'A method for the measurement of a real correspondence', *Papers of the Michigan Academy of Science, Arts and Letters* 49, 333–44.

Muller, J. C. (1967) 'Objective and subjective comparisons in choropleth mapping', *Cartographic Journal* 4, 156–66.

Olson, J. (1972) 'The effect of class interval system in choropleth map correlation', *Canadian Cartographer* 9, 44–9.

Robinson, A. H. and Bryson, R. A. (1957) 'A method for describing quantitatively the correspondence of geographical distributions', *Annals, Assoc. Amer. Geogr.* 47, 379–91.

Siegel, S. (1956) *Non-parametric Statistics for the Behavioral Sciences*, New York, McGraw-Hill.

Thomas, E. N. (1968) 'Maps of residuals from regression', in Berry, B. J. L. and Marble, D. F. (eds) *Spatial Analysis: A Reader in Statistical Geography*, Englewood Cliffs, NJ, Prentice-Hall.

Tobler, W. R. (1965) 'Computation of the correspondence of geographical patterns', *Papers, Regional Science Association* 15, 131–9.

WORKSHEET

(1) Attempt to give examples of worthwhile map comparisons of all ten types outlined in Table 7.1.

(2) Figure 5.10 and Table 7.4 give data for owner-occupied houses and two-car households in the Sheffield wards in 1971. Apply Cliff's (1970) autocorrelation approach to test whether or not the spatial difference between these distributions is statistically significant. (Use the z_1 and z_2 columns as your data.)

(3) The data used to construct Figure 7.1 are shown below. Plot a scatter graph of the relationship, calculate the

Continued on next page

product-moment correlation coefficient, and fit the regression line. Draw a map of residuals from this regression and comment on your results:

	Households	
Ward	*percentage private rented*	*percentage with more than 1.5 per room*
St Margarets	39.3	2.6
Latimer	36.2	3.7
Charnwood	34.8	3.7
Spinney Hill	34.4	5.8
Wycliffe	57.4	11.8
The Castle	53.4	4.2
Westcotes	44.7	2.1
Newton	2.8	1.0
The Abbey	12.4	0.8
Belgrave	25.5	1.5
Humberstone	5.8	1.2
Evington	6.3	0.5
Knighton	32.3	1.2
De Montfort	12.6	2.0
Aylestone	8.0	0.9
North Braunstone	4.5	3.1

Note: To identify wards, see Figure 5.2 (p. 116).

INDEX

Note: Page numbers in italic refer to major sections